Physics in Easy

© 2023 Dipl.-Ing. Matthias Badelt

Physics in Easy

Volume 2:
Mechanics for High School and College

Translation of the 2nd edition
with vocabulary list for German technical terms.

Bibliographic information of the German National Library: The German National Library lists this publication in the German National Bibliography; detailed bibliographic data are available on the Internet at dnb.dnb.de.

Bibliographic information of the German National Library: The German National Library lists this publication in the German National Bibliography; detailed bibliographic data are available on the Internet at dnb.dnb.de.

Matthias Badelt
Auf dem Rott 31
59069 Hamm
Germany
www.nachhilfe-rhynern.de

Production and publishing:

BoD - Books on Demand, Norderstedt, Germany

ISBN: 9783743179189

The list shows books to be published in the future (as of 2023)

√ = Handwritten finished version with pictures and formulas as sketches.

√√ = The graphic revision is finished and the book has been published.

Mathematics in Easy

Volume 1   - Arithmetic and Numbers

Volume 2a - Algebra √

Volume 2b - Linear Algebra

Volume 3a - Analysis: Functions with one Variable √

Volume 3b - Analysis: Function Families, Multivariate
Functions, Vector Analysis

Volume 3c - Analysis: Modeling, Transforming,
Interpolation

Volume 3d - Analysis: Mathematics of Economics

Volume 4   - Geometry and Vector Algebra √

Volume 5a - Descriptive Statistics √

Volume 5b - Evaluative Statistics

Volume 6   - Differential Equations

Volume 7a - Sequences, Series, Proof Methods

Volume 7b - Financial Mathematics √

Volume 7c - Numerics

Physics in Easy

Volume 1 - Statics and Hydrostatics √

Volume 2 - Mechanics: Kinematics, Dynamics, Energy √√

Volume 3 - Electricity Theory √

Volume 4 - Oscillations, Waves and Optics

Volume 5 - Thermodynamics

Volume 6 - Mathematics in Physical Applications

Volume 7 - Atomic Physics

Volume 8 - Relativity and Astronomy

Engineering Mechanics in Easy

Volume 1: Engineering Mechanics
for Technicians and Engineers √

I write in parallel on all books at the same time, depending on the subjects I have in my timetable. Each year, one book is to be published in German and in English.

Table of Contents

1 Fundamentals

1.1 Vocabulary

linear motion, linear movement	lineare Bewegung	to accelerate	beschleunigen
rectilinear motion	geradlienige Bewegung	speed up, become faster	schneller werden
circular motion	kreisförmige Bewegung	to slow down	langsamer werden
superimposed motion	überlagerte Bewegung	to brake	bremsen
superposition	Überlagerung, Superposition	uniform motion	gleichförmige Bewegung
kinematics	Kinematik	uniform accelerated	gleichförmig beschleunigte
dynamics	Dynamik	movement	Bewegung
conservation of energy	Energieerhaltung	centrifugal force	Zentrifugalkraft
conservation of momentum	Impulserhaltung		

- **Kinematics** is the description of the motion without considering the inertial mass (chapter 2).
- **Dynamics** is the description of the motion with consideration of the inertial mass (chapter 4).

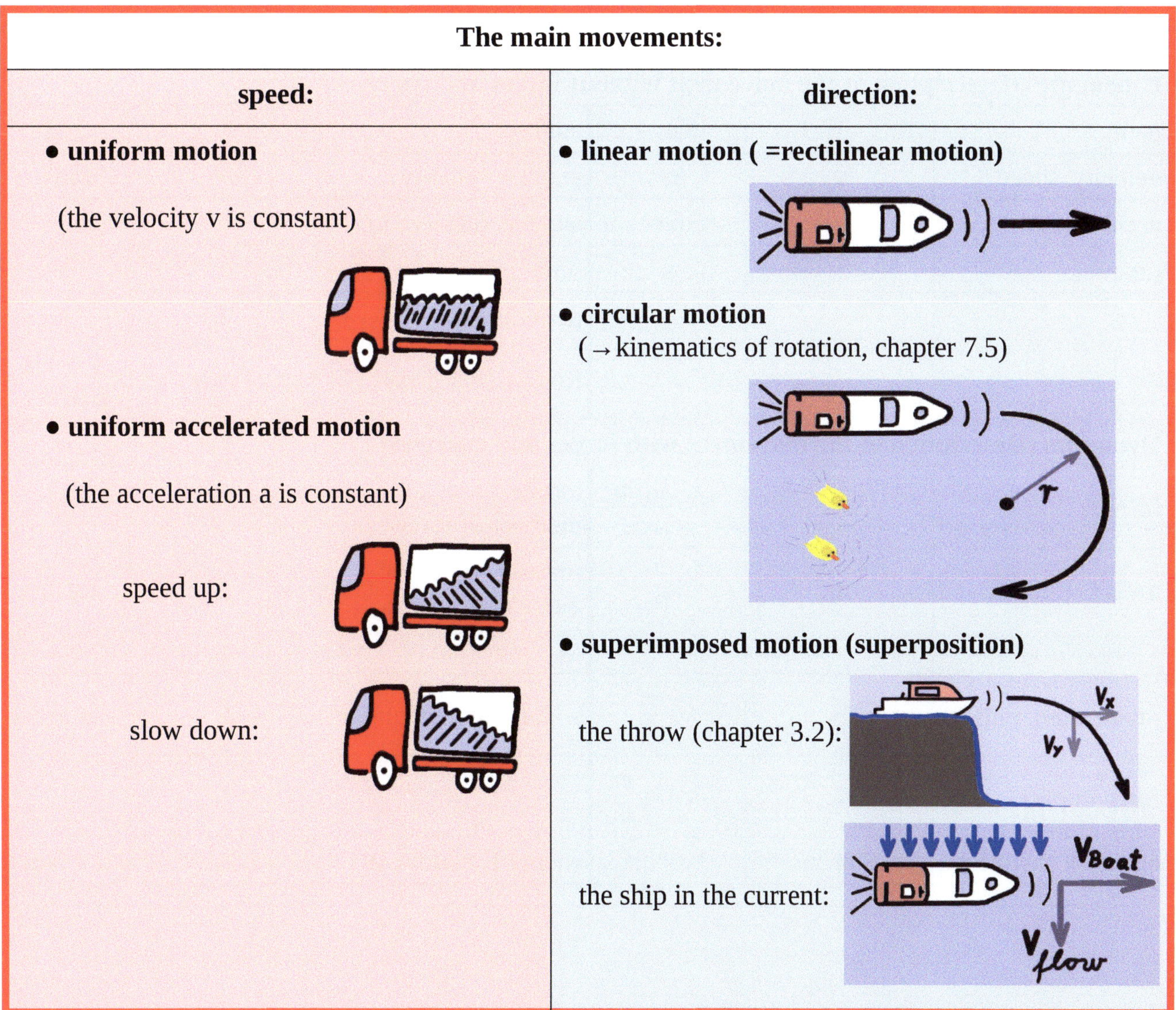

1.2 Formula Symbols and Units

translation	Translation	angle in degrees or radians	Winkel in Grad oder Radiant
rotation	Rotation	angular velocity	Winkelgeschwindigkeit
path, distance	Weg	angular acceleration	Winkelbeschleunigung
velocity, speed	Geschwindigkeit	frequency, drive	Frequenz, Drehzahl
acceleration	Beschleunigung	period, periodic time, cycle duration	Periodendauer
force	Kraft	torque	Drehmoment
mass	Masse	moment of inertia	Trägheitsmoment
momentum, pulse	Impuls	angular momentum	Drehimpuls
force impact	Kraftstoß		

The table shows an overview of the most important formula symbols that will appear in Kinematics and Dynamics:

Translational Movement (parallel displacement of a body)			Rotational Movement (spinning of a body)		
denomination	formula symbol	unit	denomination	formula symbol	unit
Kinematics (Description of the movement without forces and energies)					
path	s	m	angle	φ	rad
velocity, speed	v	m/s	angular velocity	ω	rad/s
acceleration	a	m/s²	angular acceleration	α	rad/s²
time	t	s	time	t	s
			periodic time	T	s
			frequency = drive	f or n	s^{-1}= Hz (Hertz)
Dynamics (Description of the movement with forces and energies)					
force	F	$N=\dfrac{kg \cdot m}{s^2}$	torque, moment of force	M (or τ)	Nm
mass	m	kg	(mass-) moment of inertia ($\rightarrow$ area moment of inertia I see vol. Engineering Mechanics)	J	kg m²
momentum, pulse	p	$kg\dfrac{m}{s}$	angular momentum	L	kg m²/s
force impact	Δp	$Ns=kg\dfrac{m}{s}$			
energy E = work W	E or W	J = Nm	energy E = work W	E	J = Nm

 © Badell.de

1.3 The Representation of Movements in Diagrams

uniformly moved	gleichförmig bewegt	integral	das Integral
uniformly accelerated	gleichförmig beschleunigt	to integrate	integrieren
path-time diagram	Weg-Zeit-Diagramm	primitive function	die Stammfunktion
velocity-time diagram	Geschwindigkeits-Zeit-Diagramm	detivative, derivation f'(x)	die Ableitung f'(x)
		derivation of a formula	die Herleitung einer Formel
		to derive, deriving	ableiten
		slope	die Steigung

The table shows the types of motion in the path-time diagram s(t), the velocity-time diagram v(t) and the acceleration-time diagram a(t).

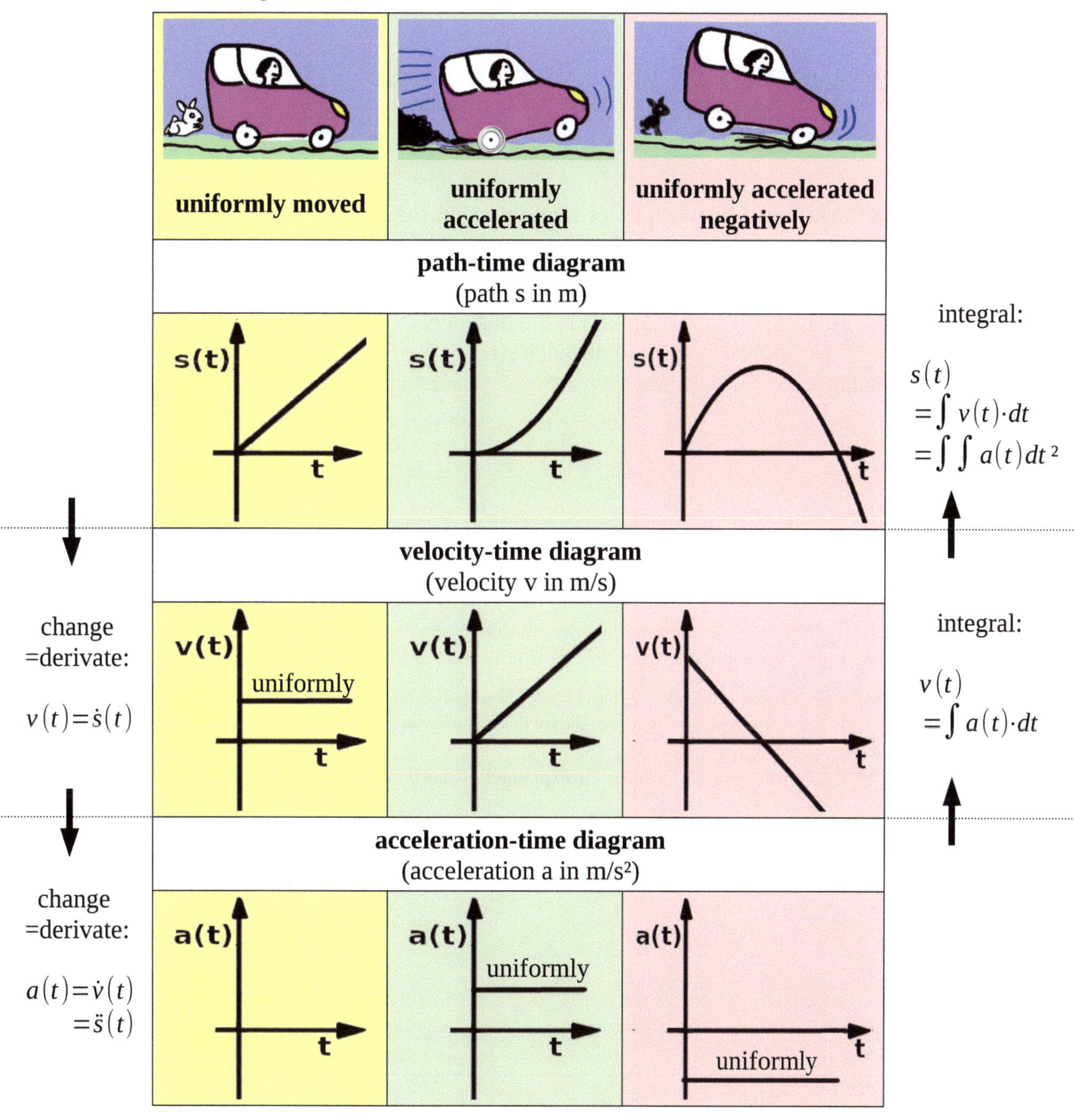

$$s(t) = \int v(t)\cdot dt = \int \int a(t) dt^2$$

integral:

$$v(t) = \int a(t)\cdot dt$$

change =derivate:

$$v(t) = \dot{s}(t)$$

change =derivate:

$$a(t) = \dot{v}(t) = \ddot{s}(t)$$

Explanation of the term "derivative" using the example of the third column (table above):

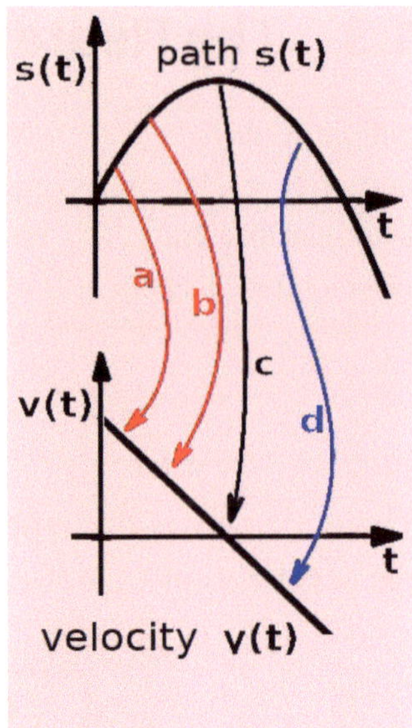

> The change of the path s(t) is the velocity v(t).
> Or: The <u>derivative</u> of the path s(t) is the velocity v(t).
> Or: The <u>derivative</u> of the function s(t) is the function s'(t) = v(t).
> Or: The slope of the function s(t) is the <u>derivative</u> s'(t) = v(t).

  a) The car is moving fast:
      The path s(t) has a large gradient. → The derivative v(t) has a large function value.
  b) The car moves more slowly:
      The distance s(t) still increases, but less. → The derivative v(t) has a smaller function value.
  c) The car stops:
      The distance s(t) no longer increases. → Derivative v(t) = 0
  d) The car moves backwards: The derivative v(t) is negative.

Explanation of the term "antiderivative" or integral using the example of the third column (table above):

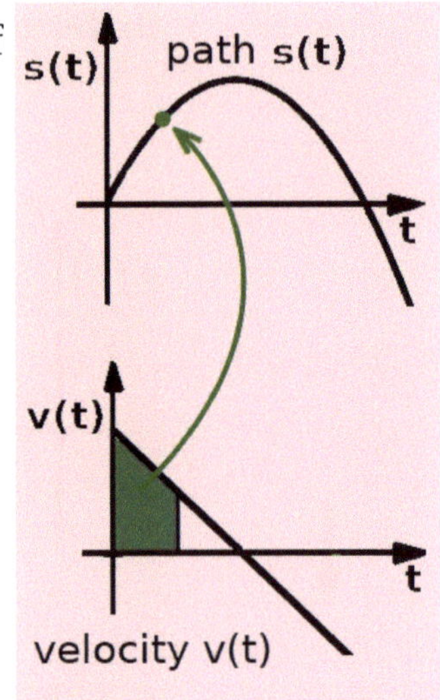

> Velocity v times time t is the distance.
> Or: The area under the velocity function v(t) is the distance.
> Or: The integral of the velocity v(t) is the distance s(t).
> Or: $\int v(t) \cdot dt = s(t)$
> (Or: $\sum v(t) \cdot \Delta t \approx s(t)$)

> The longer the car travels at a speed v(t),
> the greater the distance covered = area under the curve.

■ **Note**

The calculation of a derivative (=differential calculus) is explained in the book "Mathematics in easy, Volume 3a: Curve Sketching".

We only need these calculations here to derive our formulas (→ Chapter 2.1). <u>If you are not interested in derivations of formulas, you can work through this volume without any knowledge of differential calculus.</u>

EXERCISES:

1. The Area in the v-t Diagram

a) What is the physical significance of the area under the curve in a v-t diagram?

b) What is the result of the integral $\int_{t_1}^{t_2} v(t)\cdot dt$?

c) The figure on the right shows a v-t diagram:

- Describe the individual sections using the technical terms (uniformly moved, uniformly accelerated, uniformly negatively accelerated).

- Calculate the slope of the function v(t) and the area under the function v(t) for all three areas. Which physical unit results for the slope and for the area? What is the significance of the results?

- Draw the corresponding s-t and a-t diagrams.

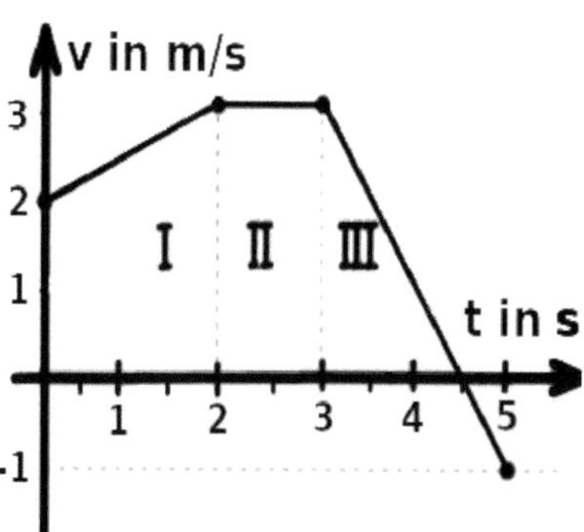

2. The Stopping Distance in the Diagram:

If a vehicle has to brake suddenly, the distance required is called the stopping distance. It is valid:

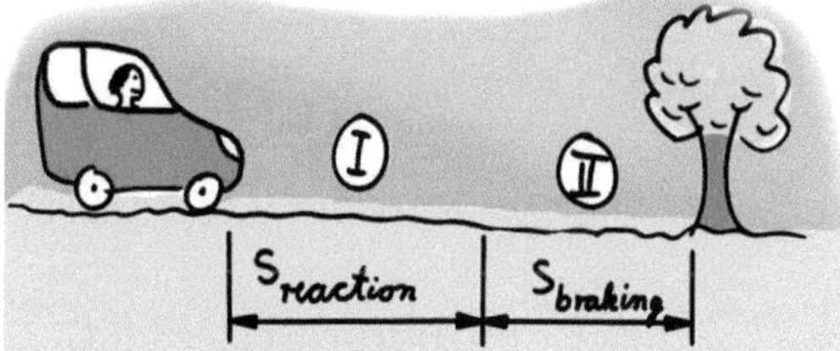

Stopping distance = reaction distance + braking distance

The brakes are not applied during the reaction time (reaction distance).

a) Sketch an s-t, a v-t and an a-t diagram for the stopping distance (principle sketch without numbers).

b) What is the significance of the area under the curve in the a-t diagram?

3. The Movement of a Stone:

A stone is thrown vertically into the air. Afterwards it falls back to the ground. Justify which diagram fits. What do the other two diagrams represent?

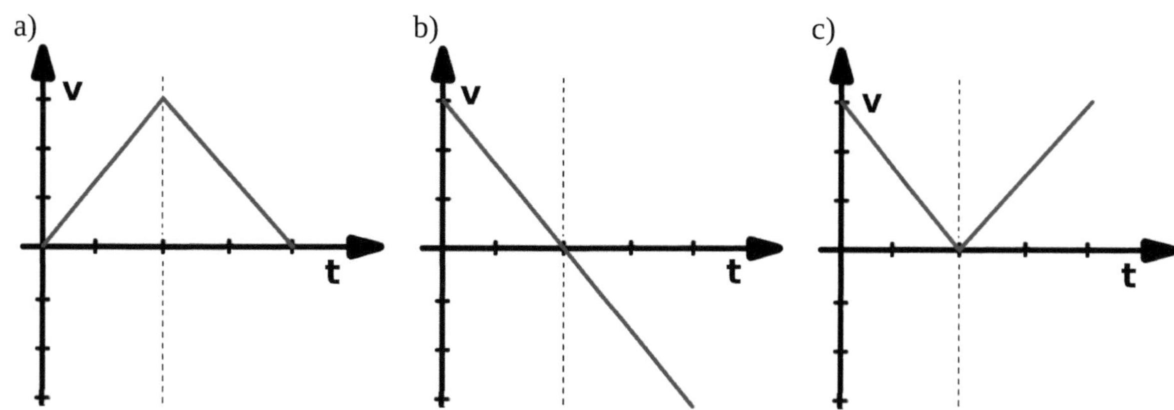

1.4 Force Directions

force application point	Kraftangriffspunkt	gravitational force F_g	Gravitationskraft F_G
center of gravity	Schwerpunkt	downhill force F_h	Hangabtriebskraft F_H
force arrow	Kraftpfeil	slope down force	Hangabtriebskraft
uniformly moved	gleichförmig bewegt	normal force F_n	Normalkraft (Andruckkraft) F_N
uniformly accelerated	gleichförmig beschleunigt	frictional force F_f	Reibungskraft F_R
slope, gradient, acclivity	Hang	inertia force F_a	Trägheitskraft F_a
uphill-slope ↔ downhill-slope	Steigung ↔ Gefälle	centripetal force F_c	Zentripetalkraft F_Z
perpendicular, at right angle	senkrecht zu.., lotrecht zu..	rigid body	der starre Körper

Drawing the arrows (see also volume "Engineering Mechanics"):
- Direction of arrow: Forces always act on a body from the outside.
- Force application point: Forces may be drawn pulling on the body or pushing on the body.

Simplification:
- In this overview, it is assumed for simplicity that all forces act on the center of gravity (actually, the friction force acts on the wheels).

forward	uniformly moved	uniformly accelerated	braking (uniformly negative accelerated)	**gravitational force (weight force) F_G: downward.** **frictional force F_R: Opposite to the velocity v**
backward	uniformly moved	uniformly (backward) accelerated	braking	**inertia force F_a: Opposite to the acceleration a**
looping				**radial forces or central forces:** **The centrifugal force: outwards**
	\• The centrifugal force arises due to the mass inertia. \• The centripetal force pulls the body to the center of the circle. It can be, for example, a rope force or a gravitational force.			**The centripetal force: inwards**
slope				**downhill force: parallel to the slope** **normal force: perpendicular to the slope**

EXERCISES

1. **Force Directions on the Slope:** A car is driving up a hill. The car accelerates.

   a) Sketch the car with the following forces: weight force, downhill force, normal force, inertia force, friction force.

   b) Which force changes its direction when the car brakes?

   c) In your sketch, use the right angles to see the relationship between the downslope force, weight force, and normal force. What is the normal force if the car has a weight force of 11 kN and the slope down force is 1.2 kN?

2. **Force Directions on the Rope:** A ball is hanging from a rope on a ceiling. It is pushed so that it moves on a circular path. The angle of the rope to the vertical is called α.

   a) Draw all the forces acting on the ball: In which direction does the centrifugal force act? In which direction does the rope force act? In which direction does the weight force act?

   b) In horizontal direction: The centrifugal force is opposite to the horizontal rope force component $F_{sx} = F_s \cdot \sin(\alpha)$. Write this down as an equation.

   c) In the vertical direction: The weight force is opposite the vertical rope force component F_{sy} . Write this down as an equation.

   d) In the direction of the rope: Centrifugal force and weight force can be added to a common force in the direction of the rope via Pythagoras. This is opposed by the rope force. Write this down as an equation.

   e) Enter the three directions (vertical, horizontal and rope direction) in your drawing and assign the three directions to the three equations.

2 Kinematics of Translation

path, distance	Weg	to seek, sought	suchen, gesucht
velocity, speed	Geschwindigkeit	equations of motion	die Bewegungsgleichungen
acceleration	Beschleunigung	time-free form	zeitfreie Darstellung

Term: **Kinematics** deals with the movement of bodies. A distinction is made between translation and rotation (chapter 6). Kinematics does not deal with forces and inertial masses. This topic is assigned to dynamics.

2.1 The Equations of Motion

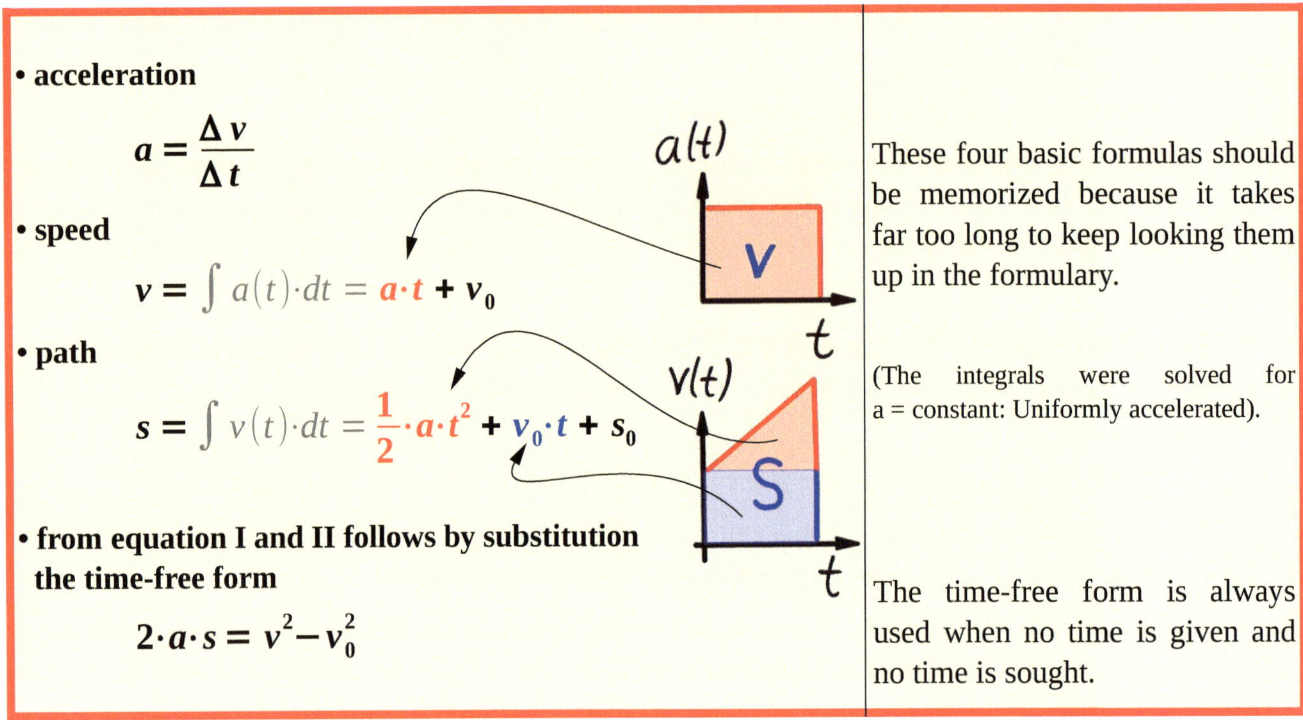

- **acceleration**

$$a = \frac{\Delta v}{\Delta t}$$

- **speed**

$$v = \int a(t)\cdot dt = a\cdot t + v_0$$

- **path**

$$s = \int v(t)\cdot dt = \frac{1}{2}\cdot a\cdot t^2 + v_0\cdot t + s_0$$

- **from equation I and II follows by substitution the time-free form**

$$2\cdot a\cdot s = v^2 - v_0^2$$

These four basic formulas should be memorized because it takes far too long to keep looking them up in the formulary.

(The integrals were solved for a = constant: Uniformly accelerated).

The time-free form is always used when no time is given and no time is sought.

Table 2.1: Equations of Motion

We need the integral (chapter 1.3) only for derivations. Tasks for integrals see Volume 6: Mathematics in physical applications.

2.2 Exercises to get started

lead, head start	Vorsprung	path, distance	der Weg
the distance covered	die zurückgelegte Strecke	velocity, speed	die Geschwindigkeit
		acceleration	die Beschleunigung

Formula used:

$$s = \frac{1}{2} \cdot a \cdot t^2 + v_0 \cdot t + s_0$$

a = acceleration
t = time
v = velocity
s = distance or path

1. **Distance Covered in three Minutes:** An athlete runs for three minutes at a time. What distance does he cover?

	Solution hint:	Solution:
a) Its speed is 2 m/s	$a=0; \quad v_0 = 2\frac{m}{s}; \quad s_0 = 0\,m$	360 m
b) Its speed is 2 m/s and it has a lead of $s_0 = 500$ m.	$a=0; \quad v_0 = 2\frac{m}{s}; \quad s_0 = 500\,m$	860 m
c) His running speed is initially 2 m/s and he has a head start of 500 meters. In addition, he becomes slower and slower: Braking acceleration $a = -0,01\,m/s^2$	$a=-0,01\frac{m}{s^2}; \quad v_0 = 2\frac{m}{s}; \quad s_0 = 500\,m$	698 m
d) He starts with 0 m/s and no lead. His acceleration is $a = +0,01\,m/s^2$	$a=+0,01\frac{m}{s^2}; \quad v_0 = 0\frac{m}{s}; \quad s_0 = 0\,m$	162 m

2.3 Exercises for Uniform Movement

conveyor, conveyor belt	das Fließband	uniformly moved	gleichförmig bewegt

Formulas used:

$$s = \left(\cancel{\tfrac{1}{2} \cdot a \cdot t^2} \right) + v_0 \cdot t + s_0$$

$$v = \left(\cancel{a \cdot t} \right) + v_0$$

Moving uniformly means: Acceleration a = 0. Accordingly, the formula simplifies.

1. **Uniformly moving Conveyors:** A piece of luggage is transported over two conveyor belts in the airport. The first conveyor belt is 200 m long and has a speed of 10.8 km/h (= 3m/s). The second conveyor belt is 20 m long. For this conveyor belt the piece of luggage needs 16 seconds.

   a) What time does the piece of luggage need on the first conveyor belt?

   b) What speed does the piece of luggage reach on the second conveyor belt?

   c) What is the average speed for the total distance?

   d) Draw the v-t diagram and the s-t diagram for the total distance!

2. **Meeting Point of two Uniformly Moving Vehicles (1v2):** The cities of Hamm and Soest are 50 km apart. At time t = 0 a cyclist sets off from Hamm to Soest. He has the speed v = 5 m/s. At the same time, a motorcyclist sets off in the opposite direction in the town of Soest. He has a speed of v = 72 km/h.

   a) Set up the distance-time law s(t) for both vehicles.

   b) When and where do the two meet?

3. **Meeting Point of two Uniformly Moving Vehicles (2v2):** The distance between location A and location B is unknown. At 9 o'clock a cyclist sets off from place A to place B. His speed is v = 5m/s. Ten minutes later, a motorcycle (v = 72 km/h) travels in the same direction.

   a) Make an s-t diagram (without numerical values).

   b) At what time do they meet?

   c) What distance do they cover until the cyclist is caught up?

4. **Average Speed during Interval Jogging:** Greta ruined her knees due to long winter breaks and false athletic ambition during the vacations. Her doctor knows that cartilage has no veins and relies on regular, gentle exercise for nourishment via synovial fluid. In hopes of long-term cartilage rebuilding, Greta now takes it upon herself to jog slowly and gently with barefoot shoes on the ball of her foot and to switch immediately to gentler walking when pain occurs. She also wants to take a break for one or two days after each workout to regenerate micro-injuries.

   After surviving the first muscle soreness due to the change in running style, Greta, out of curiosity, comes up with the idea of analyzing her workout data, which she - as a privacy-conscious data protector - has of course recorded using an app from the F-Droid store. She reads out the data below.

Calculate the average speed in each case:

a) In the first months of training, Greta runs at 4 km/h for 70 minutes, 8 km/h for 30 minutes, and 10 km/h for 8 minutes.

b) Later, Greta usually runs five kilometers at 8 km/h, four kilometers at 4 km/h, and one kilometer at 10 km/h.

c) After one year she runs six kilometers in 45 minutes and four kilometers at 5 km/h.

d) Has she continuously improved her fitness (speed)?

2.4 Exercises for Uniform Acceleration

Formulas used:

$$s = \frac{1}{2} \cdot a \cdot t^2 + v_0 \cdot t + s_0$$
$$v = a \cdot t + v_0$$
$$a = \frac{\Delta v}{\Delta t}$$

1. **Free Fall:** A body falls from a height of 10m (acceleration due to gravity $a_{earth} = g = 9{,}81\frac{m}{s^2} = 9{,}81\frac{N}{kg}$). There are three different ways to draw a suitable coordinate system:

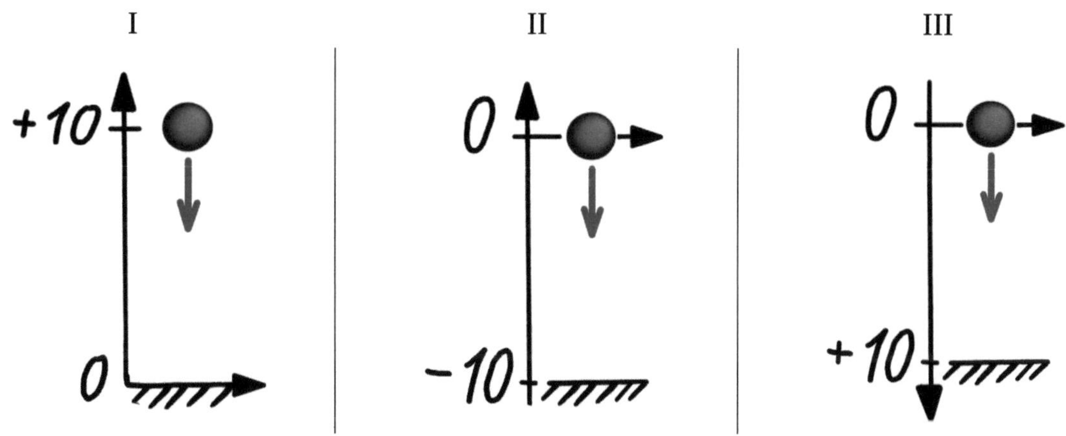

a) How long does the ball fall? Make a separate calculation for each of the three coordinate systems. Which of the calculation methods is easier?

b) What is the velocity at impact?

c) Draw the diagrams s(t), v(t) and a(t) for each of the three coordinate systems.

2. **Vertical Throw:** A ball is thrown vertically upwards with a velocity of 15 m/s. The ball is thrown vertically downwards.

a) How long is the ball in the air before it reaches the bottom again?

b) How high does it fly?

c) At what height is the ball after one second, after two seconds and after three seconds?

3. **Braking Distance of a Cyclist:** A cyclist brakes with a braking acceleration of - 2 m/s².

   a) What time does he need to brake from 30 km/h to 20 km/h?

   b) What distance does he cover in the process?

2.5   Exercises without Times

Formulas used:

$$2 \cdot a \cdot s = v^2 - v_0^2$$

1. **Acceleration and Braking Distance of a Motorcycle:** A motorcyclist accelerates from 100 km/h to 150 km/h on a 500m long track.

   a) How great is his acceleration?

   b) He then brakes because an obstacle appears 300m away. His braking acceleration is a = -3 m/s². How far in front of the obstacle does he stop?

2. **Overtaking a Racing Car:** A racing car is standing on a lane. A competitor in a second racing car passes at 3 m/s. From that moment both cars accelerate:

   • Race car 1 accelerates to 20 m/s within 40 meters.

   • Race car 2 accelerates constantly at 3.5 m/s².

   a) Calculate the acceleration of car 1.

   b) Calculate the time at which both vehicles have the same speed.

   c) Calculate the time at which the first vehicle has caught up with the second vehicle.

   d) Calculate the location where the first vehicle caught up with the second vehicle.

   e) Draw the v-t diagram and the s-t diagram for both vehicles.

2.6 Exercises for Sectional Movements

acceleration	Beschleunigung	reaction distance	Reaktionsweg
deceleration, slowdown	Verlangsamung	braking distance	Bremsweg
to brake	bremsen	stopping distance	Anhalteweg

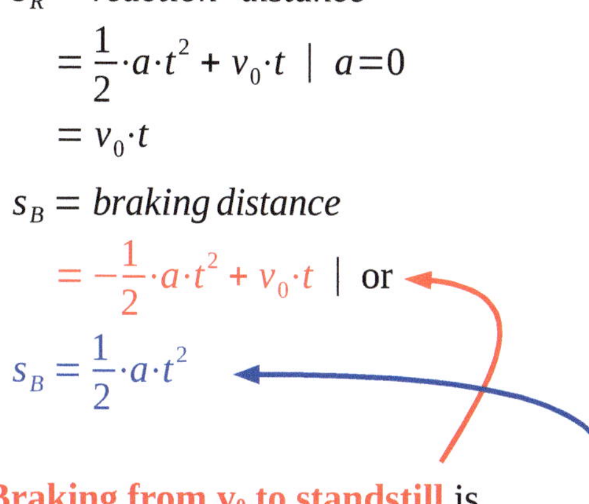

$$s_R = reaction\ distance$$
$$= \frac{1}{2}\cdot a\cdot t^2 + v_0\cdot t \mid a=0$$
$$= v_0\cdot t$$

$$s_B = braking\ distance$$
$$= -\frac{1}{2}\cdot a\cdot t^2 + v_0\cdot t \mid \text{or}$$
$$s_B = \frac{1}{2}\cdot a\cdot t^2$$

A combination of uniform motion and uniformly accelerated motion is the stopping distance of a car.

Braking from v_0 to standstill is physically identical with **acceleration from standstill to v_0.**

1. **Stopping Distance for a Passenger Car:** A passenger car is traveling at a speed of 90 km/h when the driver detects an obstacle 80 m away. Can the car stop in time if the driver reacts after 0.5 s and the braking acceleration is -5 m/s²?

2. **Acceleration Phases and Stopping Distance of a Train:** a train accelerates uniformly from standstill to a speed of 108 km/h within 60s. In another 120s it accelerates, also uniformly, to 180 km/h. After that, it runs at a constant speed for three minutes.

   a) What are the accelerations in the three phases?

   b) What is the total distance traveled?

   c) Draw the v-t and the a-t diagram.

   d) Suddenly an obstacle appears 3000 m in front of the train. The train is still traveling 180 km/h. Despite a reaction time of 1s, the train just comes to a stop. What is its braking acceleration?

   e) Another train is traveling at 200 km/h. What is the speed of this train at the obstacle? (Reaction time and braking acceleration are identical with the first train).

3. **Sectional Uniform Motion of a Passenger Car**

   a) A passenger car accelerates uniformly from 0 to 50 km/h in 10s. What distance has it covered?

   b) At the highway entrance the car accelerates further from 50 km/h to 130 km/h. The distance is 1000 m. What is the acceleration in m/s^2?

   c) An end of a traffic jam blocks the way. The driver has a reaction time of 0.3 s. The braking acceleration of the vehicle is 5 m/s^2. How long are the reaction distance, braking distance and stopping distance at $v_0 = 130$ km/h? (How long would the paths be at $v_0 = 250$ km / h?)

   d) Two soccer fans drive 130 km from Cologne to the BVB stadium in Dortmund. The first car driver drives comfortably at a constant 120 km/h. The second car driver drives as fast as he can: 10% of the way with 190 km/h, 40% of the way with 150 km/h and 50% of the way with 120 km/h. Calculate the travel times for both drivers.

4. **Stopping Distance with unknown Initial Speed and Moose (difficult):** The Arctic Circle, like the equator, runs all the way around the Earth. It marks the place from which there is no sunlight for at least one day in winter. Due to climate change, even here temperatures of 30 degrees in summer have already been reached. Moose then like to hide in cool road tunnels. Therefore, on hot summer days, serious traffic accidents with moose occur here again and again.

A car driver enjoys the vacation on a Norwegian country road, when he suddenly sees a moose in the tunnel 70m in front of him. His reaction time is 0.4s. The braking acceleration of the vehicle is -5m/s^2. What is the maximum permissible initial speed of the car so that he can brake in time to avoid the moose?

3 Kinematics of Translation - The Superimposed Movement (=Superposition)

3.1 Examples for Superimposed Movements

superposition	Superposition, Überlagerung	(river) current, flow	Strömung
superimposed movement	überlagerte Bewegung		

■ **Superposition (Addition of Movements):**

The speed of the person is superimposed on the speed of the train.

The speed of the water is superimposed on the speed of the boat.

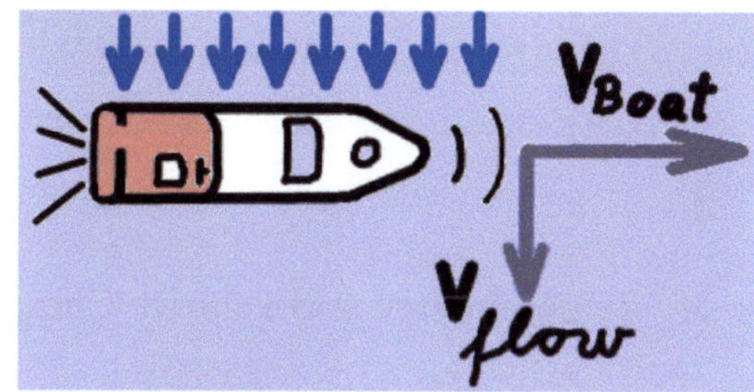

The uniform motion in x-direction overlaps with the accelerated motion in y-direction.

3.2 Superimposed Movement - The Throw

superposition	Superposition, Überlagerung	(river) current, flow	Strömung
x-direction, y-direction	x-Richtung, y-Richtung	maximum, peak	Hochpunkt, Maximum
fall time, falling time, time of fall	Fallzeit	highest point	Hochpunkt, Maximum
flight time, flying time, time in the air	Flugzeit	point of impact, strike point	Auftreffpunkt
		landing point	Landepunkt

■ **Symmetries and Times in Throwing Tasks**

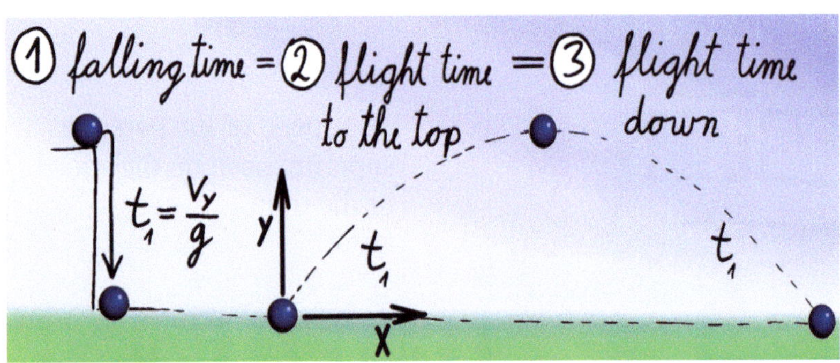

The ball does not care whether it falls from the highest point or is thrown to the highest point:

The time required for the y-path is always the same.

■ **Trigonometry and Velocity Components in Throwing Tasks**

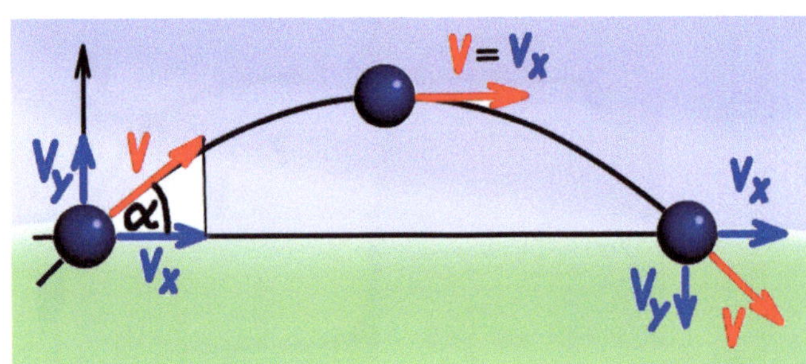

Trigonometry: In the triangle is valid:

$$v_x = v\cdot\cos(\alpha) \qquad v_y = v\cdot\sin(\alpha)$$

$$\tan(\alpha) = \frac{v_y}{v_x}$$

Pythagoras is also valid:

$$v^2 = v_x{}^2 + v_y{}^2$$

■ **The Maximum Throwing Distance**

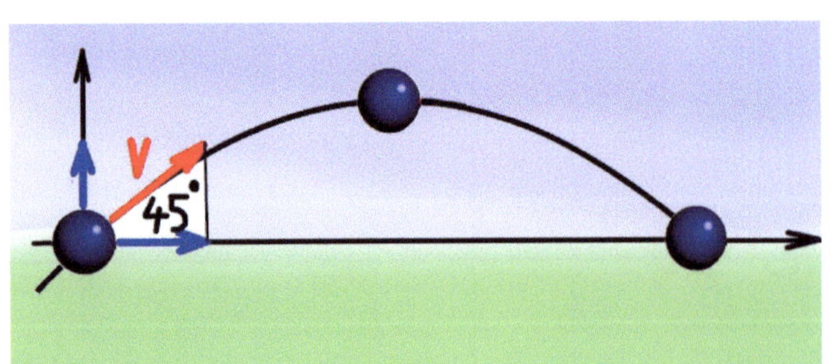

The maximum throwing distance is achieved at a throwing angle of 45°.

Derivation: First calculate the time.

$$s_y(t)=0=-\frac{1}{2}\cdot g\cdot t^2+v_{0y}\cdot t \Leftrightarrow t=\frac{2v_{0y}}{g}$$

Then insert in $s_x=v_0\cdot x\cdot t \Rightarrow s_x(\alpha)=..$
Maximum: Set derivative = 0.

 Badelt.de

■ Formulas for Throwing Tasks (Equations of Motion)

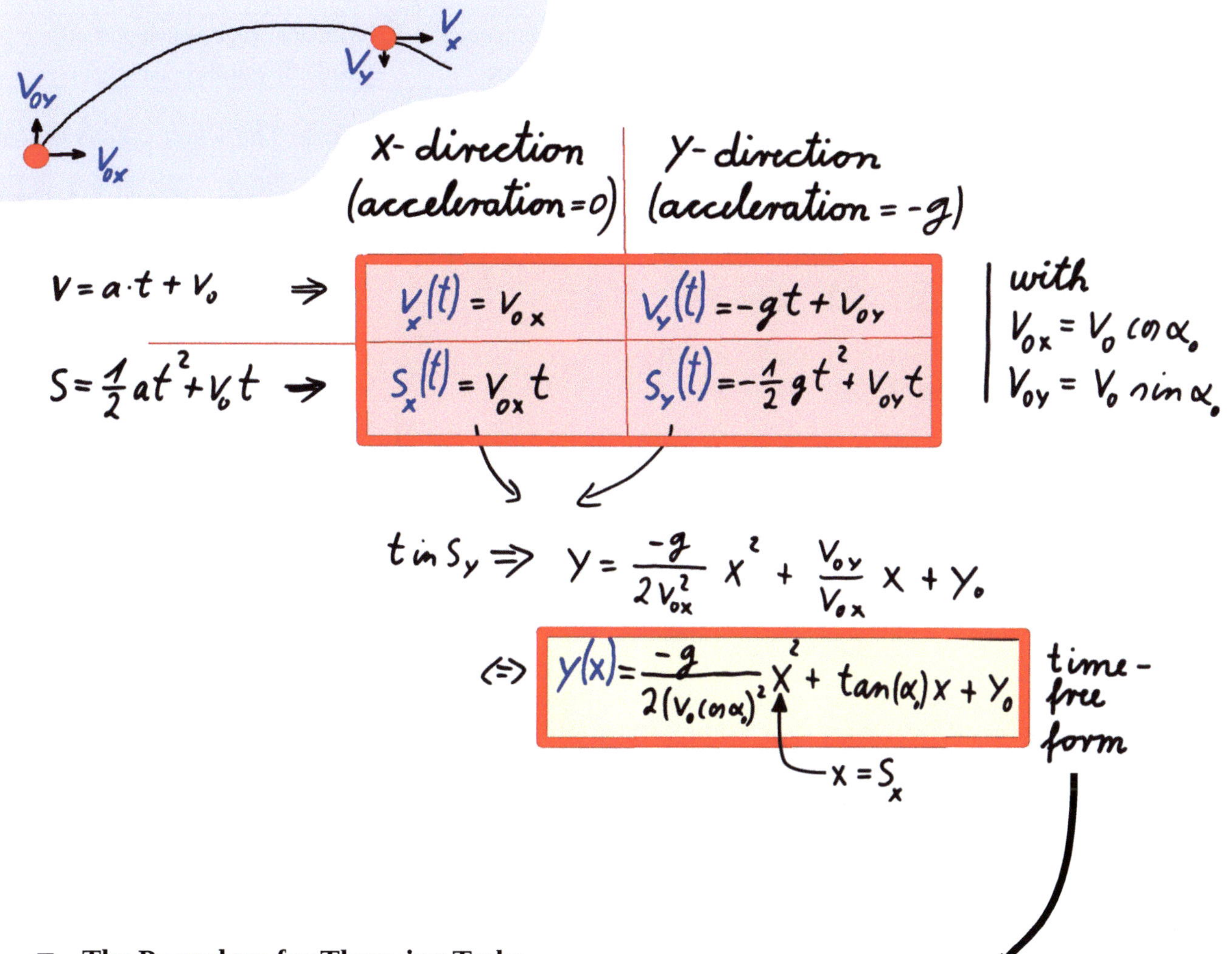

■ The Procedure for Throwing Tasks

With Calculation of Time	With the Use of the Time-free Representation
• Always consider the movements in x-direction $s_x(t)$, $v_x(t)$ and in y-direction $s_y(t)$, $v_y(t)$ separately. • **Always calculate the time t first** • Basically, in physics, write down all formulas (on the subject). For each formula, consider how many unknown variables it contains. Is there a formula with only one unknown? Or are there two formulas with a total of 2 unknowns?	• If the time is neither given, nor searched for, then use the time-free representation: $$y(x) = \tan(\alpha_0) \cdot x - \frac{g}{2v_0^2 \cos^2(\alpha_0)} \cdot x^2$$ • Or when considering the y- direction alone: $$2 \cdot g \cdot s = v^2 - v_0^2$$

3.3 Exercises

the horizontal throw	der waaregechte Wurf	velocity components	Geschwindigkeitskomponenten
the oblique throw	der schiefe Wurf	angle of impact	Auftreffwinkel

1. **Horizontal Throw:** Peter wants to water a bed with his garden hose. The water comes out of the hose horizontally at a speed of 18 km/h. The bed is 3 m away. The bed is 3 m away.

   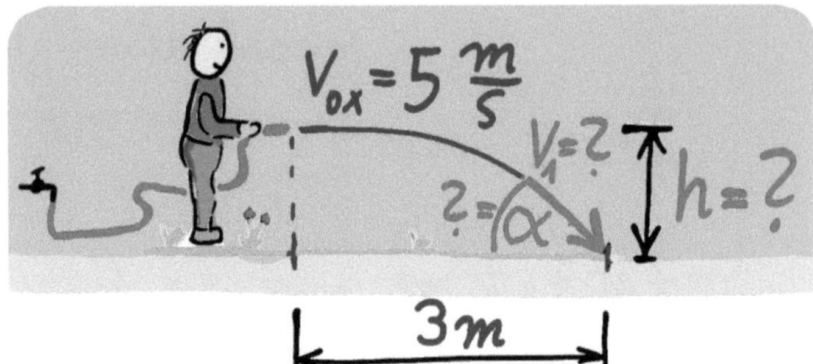

   a) How high must he hold the hose?

   b) What are the angle of impact and the speed of impact?

2. **Velocity Components and Angle during Horizontal Throw:** After catastrophes (e.g. flood, earthquake, war...) it can happen that relief packages are dropped from airplanes. Also at the time of the Berlin Airlift (1948-1949) small packages were dropped from time to time (raisin bombers).

   [Historical background: At the time of the cold war between the Eastern bloc (Russia, East Germany...) and the Western powers (USA, France, Great Britain), Russia tried to annex West Berlin, which belonged to the Western powers. In order to take Berlin, all roads and supply routes from the West to West Berlin were blocked. As a result, the West began to use airplanes to fly 2000 tons of food, coal, gasoline and medicine to West Berlin every day].

   An airplane flies 200m high and drops small packages. The initial speed is 170 km/h. (For all packages; horizontal drop; the parachute does not open).

   a) Calculate the impact velocity and the impact angle (to the horizontal) on the ground.

   b) Velocity components: The aircraft is now flying at the same speed at a different altitude. A second package impacts at a speed of 330 km/h. Calculate at impact: v_y and α (for parcel II).

   c) Calculate the (new) drop height and the drop duration (for parcel II).

   d) A third parcel has an impact angle of $\alpha = 50°$ (to the horizontal). The impact velocity was 70 m/s. What were the magnitudes of v_x and v_y at impact (parcel III)?

3. **Find the Appropriate Formulas (Task without Numerical Values):** A ball rolls over a table with a velocity v_0. The table has a height h above the ground.

   a) What is the velocity v_x of the ball when it hits the ground?

   b) What is the velocity v_y of the ball when it hits the ground?

   c) What is the velocity v of the ball when it hits the ground?

   d) At what angle α does the ball hit the ground?

4. **Oblique Throw (and Volume Flow):** Peter wants to water his bed again. Now he holds the hose up at an angle so that the water hits the flowers at the same height some distance away.

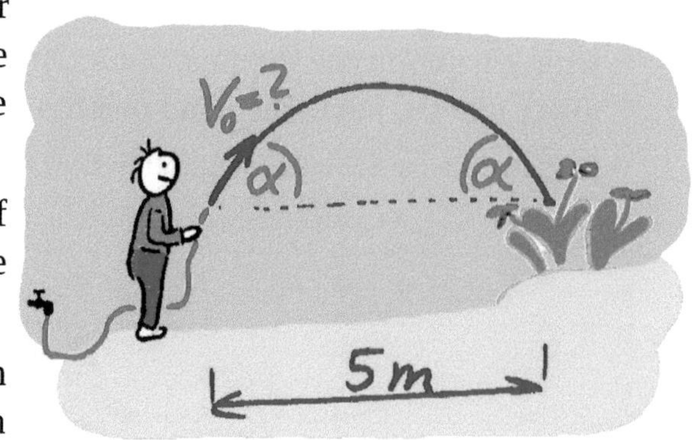

a) The water jet reaches a maximum distance of 5m. At what angle is the maximum range always achieved in principle?

b) At what speed does the water emerge from the hose? (Use the time-free representation of the equation of motion for the calculation).

c) How many liters of water per minute flow through the hose if it is a ¾ inch hose? (1 inch = 25.4 mm).

5. **Oblique Throw:** In the shot put, a ball is thrown at an optimal angle at a height of 1.80 m. The ball has an initial velocity of 1.5 mm. The ball has an initial velocity of v =40 km/h. Let the origin of the coordinate system be the throwing point.

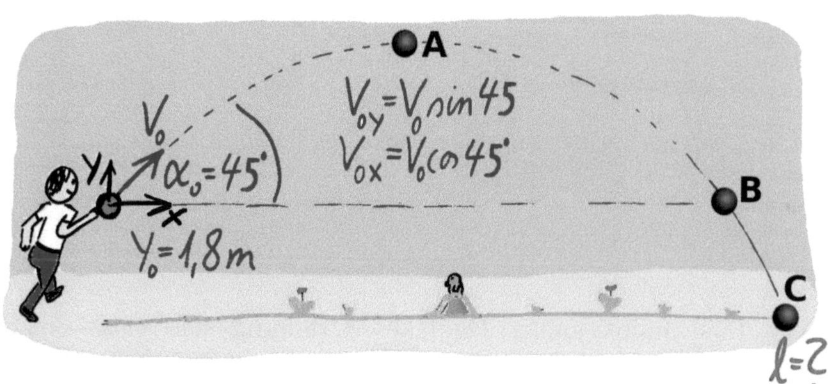

a) Set up the equations of motion for sx(t) and sy(t).

b) What is the maximum height of the ball (point A)?

   Hint: Consider symmetries. The ball could also just fall from point A to the drop height. This is the same time and the same y-distance. Calculate the time first, then the path.

c) The ball lands at point C at -1.8 meters. How far does it fly?

   Hint: Either use the time-free representation for throwing parabolas. The ball lands at sy=-1.8m. Or you calculate the fall time for the fall from A to C. Then calculate the path sx from O to C.

d) How high is the ball after two meters of flight (s_x = 2m)?

   Hint: Use the time-free form of the trajectory.

6. **Superimposed Motions:** A river (3 m/s) is swum through by a swimmer (2 m/s). The float moves exactly perpendicular to the direction of the flow.

   How far is the float carried away by the current if the river is 50 m wide?

7. **Air Hamster:** A Lufthamsda (= flying hamster) is blown off with a blowpipe (empty toilet paper roll) at an angle of 25° (v_0= 2m/s). How long and how far does it fly when the landing pillow is at launch height?

8. **Additional Task with Extreme Value Problem (difficult):** Julius and Vincent have come up with a game in the outdoor pool. On a hot summer day, one of them jumps from the diving tower into the pool. The other one has to try to hit him in flight with a ball.

(see also Volume 6: Mathematics in Physical Applications)

First Julius jumps. He takes a running start (v_{V0}= 3m/s) and jumps horizontally from the position (0| 6) to the right. At the same time Vincent, who is swimming in the water, throws a ball at an angle of 30° to the left from the position (7| 0) ($v_{B,0}$ = 8m/s). The position specifications refer in each case to the center of gravity of the bodies.

a) Calculate the smallest distance that the centers of gravity of both bodies reach in flight.

b) Simplifying, we assume that Julius has the shape of a sphere with a diameter of 140 cm. The ball has a diameter of 30 cm. Will Julius be hit by the ball in flight?

Hint: Extreme value problem:
- Main condition: The distance is to be minimized $c=\sqrt{a^2+b^2}$ (Pythagoras)
- Constraint: a(t)=... b(t)=...
- Set up objecitve function c(t)=...
- Calculate extreme values
  - necessary condition $\dot{c}(t)=0$ or d/dt (radicand) = 0
  - sufficient condition $\ddot{c}(t)\neq 0$ (dot = derived according to the time)
  - boundary extrema c(0)=... c(∞)=...

4 Dynamics of Translation

4.1 Terms

statics	Statik	stationary system	unbewegtes System
kinematics	Kinematik	inertial mass	träge Masse
dynamics	Dynamik	accelerate / decelerate	beschleunigen / bremsen

statics
(see volume „statics")
(see vol. „engineering mechanics)
= description of forces in stationary systems
$$(\quad \sum F_i = 0 \quad)$$

kinematics
(see chapter 2)
= description of the movement
without consideration of the inertial mass
$$(\quad v = a \cdot t + v_0 \qquad s = \frac{1}{2} \cdot a \cdot t^2 + v_0 t + s_0 \quad)$$

dynamics
(see chapter 4)
= description of the movement
with consideration of the inertial mass
$$(\quad F = m \cdot a \qquad W_{kin} = \frac{1}{2} \cdot m \cdot v^2 \qquad p = m \cdot v \quad)$$

4.2 Force and Acceleration: Newton's Laws

1st Newton's Law (statics)	A body does not accelerate (or decelerate) if no force acts on it, or if the sum of all forces acting on it is zero: $$\sum \vec{F}_i = 0$$
2nd Newton's Law (dynamics)	The sum of all forces equals mass times acceleration: $$\sum \vec{F}_i = m \cdot \vec{a} = m \cdot \dot{\vec{v}} = \dot{\vec{p}}$$ = Spelling of Leonhard Euler. In the original Newton formulated: $\sum F_i \sim \dot{\vec{v}}$ **Hint:** Momentum p see chapter 5.5 . 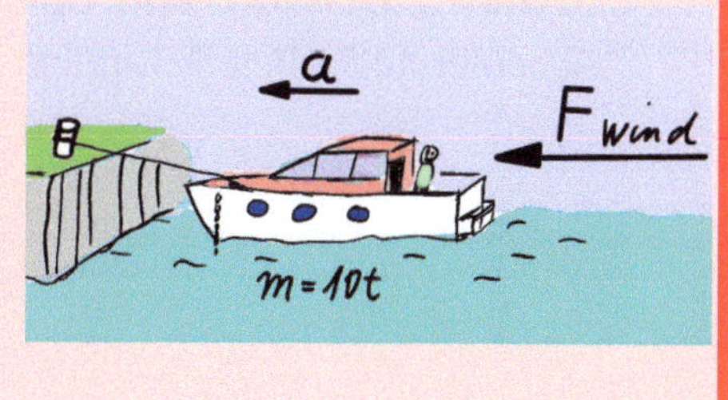
3rd Newton's Law	Force = counterforce: If a body A exerts a force on a body B, then B also exerts the same force on A.

Table 4.1: Newton's Laws

EXAMPLE:

A 10-ton motor yacht is accelerated by a gust of wind (F = 600N) at its mooring.

<ol type="a">
<li>What is the acceleration and what speed does the boat reach after 10 seconds? (Solution: 0.6m/s)</li>
<li>The boat is tied to the jetty and is now abruptly braked. How large is the braking acceleration when the boat comes to a stop after 5 cm?</li>
<li>What force acts on the rope? What mass would this force correspond to (at g = 9.81 N/kg)?</li>
</ol>

a) $F = m \cdot a$

$$\Leftrightarrow a = \frac{F}{m} = \frac{600N}{10\,000\,kg} = 0,06\,\frac{m}{s^2}$$

$$V = a \cdot t = 0,06\,\frac{m}{s^2} \cdot 10s = 0,6\,\frac{m}{s}$$

b) time free form

$$2as = v^2 - v_0^2$$

$$\Leftrightarrow a = \frac{v^2}{2s} = \frac{(0,6\,\frac{m}{s})^2}{2 \cdot 0,05\,m} = 3,6\,\frac{m}{s^2}$$

c) $F = m \cdot a = 10000\,kg \cdot 3,6\,\frac{m}{s^2} = 36\,kN$

36 kN correspond to

$$m = \frac{36000N}{9,81\,\frac{N}{kg}} = 3,67\,Tons$$

EXERCISES

1. **Known Acceleration:** A truck has a mass of 30 tons. With what force is the truck accelerated if its acceleration is a = 0.8 m/s²?

2. **Known Time and Speed:** A truck has a mass of 30 tons. What force is used to accelerate the truck if it wants to reach its speed of 80 km/h on a highway onramp within 20 seconds?

3. **Known Distance and Speed:** a truck has a mass of 30 tons. With what force is the truck accelerated when it wants to reach its speed of 80 km/h on a freeway slip road (300 m)?

4.3   The Gravitational Force

mass (in kilograms)	Masse (in Kilogramm)	gravitational field strength	Gravitationsfeldstärke
weight force (in Newton)	Gewichtskraft (in Newton)	acceleration due to gravity	Fallbeschleunigung
		acceleration due to gravity	Erdbeschleunigung

Earth	**Moon**

Backpack: <u>Mass</u> $m = 20\,kg$

<u>Weight force</u>

$$F_G = m \cdot g = 20\,kg \cdot 9{,}81\,\frac{m}{s^2} = 196{,}2\,N$$

Backpack: <u>Mass</u> $m = 20\,kg$

<u>Weight force</u>

$$F_G = m \cdot g = 20\,kg \cdot 1{,}62\,\frac{m}{s^2} = 32{,}4\,N$$

A backpack with a mass of 20 kg still has a mass of 20 kg on the moon. But on Earth, it is pulled with less force. **The 20 kg feels lighter on the moon.**

Term „weight": In scientific linguistic usage, "weight" is understood to mean the force of weight. In colloquial language, the word is also used for mass.

weight force $F_G \ = \ m{\cdot}g$	F_G = weight force in N m = mass in kg g = gravitational field strength (= acceleration due to gravity)

Earth:	Mars:	Sun:	Moon:	Pluto:
$g = 9{,}81\,m/s^2$	$g = 3{,}69\,m/s^2$	$g = 274\,m/s^2$	$g = 1{,}62\,m/s^2$	$g = 0{,}62\,m/s^2$

EXERCISES

1. Gravitational Force and Gravitational Acceleration

   a) An apple (m =200 g) falls to the ground. With which force is the apple accelerated towards the earth?

   b) An apple (m =200 g) falls to the ground. What is its acceleration?

   c) Explain the difference between a mass and a weight force.

   d) What is the mass of a passenger car when it is pushed to the ground with a weight force of 15 kN?

4.4 Frictional Force

frictional force, friction force F_f	Reibungskraft F_R	static friction F_f	Haftreibung F_R
(dry) friction	Reibung	coefficient of static friction μ_s	Haftreibungszahl μ_H
fluid friction	Flüssigkeitsreibung	kinetic friction (dynamic friction) F_f	Gleitreibung F_R
to brake, to decelerate	bremsen	coefficient of kinetic friction μ_k	Gleitreibungszahl μ_G
anti-lock braking system	Antiblockiersystem	rolling resistance (rolling friction) F_f	Rollreibung F_R
traction, tractive force F_t	Zugkraft F_Z	coefficient of rolling resistance μ_{rr}	Rollreibungszahl μ_{RR}

The friction force (or frictional force) always acts in the opposite direction to the direction of motion. Frictional force therefore leads to negative acceleration (braking acceleration).

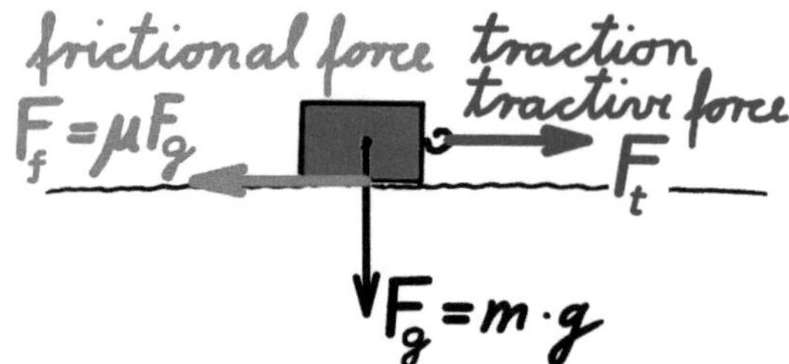

A distinction is made between static friction and sliding friction:			
static friction (body at rest)	coefficient of static friction	μ_s	$F_f = \mu_s \cdot F_n$
sliding friction (body in motion)	coefficient of kinetic friction	μ_k	$F_f = \mu_k \cdot F_n$
In addition, there is also rolling friction:			
rolling resistance (rolling friction)	coefficient of rolling resistance	μ_{rr}	$F_f = \mu_{rr} \cdot F_n$

Table 4.2: Frictional Forces

EXERCISES:

1. **Vehicle with and without anti-lock Braking System (ABS):** During emergency braking, a car (1500 kg) with locked wheels (coefficient of sliding friction μ_k = 0.6) slides toward an intersection 15 m away. The speeds are: v_0 = 50km/h and v_1 = 0 km/h.

    a) How large is the braking force?

    b) How large is the braking acceleration?

    c) Does the vehicle slide onto the intersection?

    d) The vehicle's brakes now have an anti-lock braking system. This system brakes just enough to keep the tires from slipping. So the static friction coefficient μ_s = 0.8 now applies. By how many meters is the braking distance reduced compared to the vehicle without an antilock braking system?

2. **Rolling Distance of a Ball:** A ball (m = 0.5 kg) rolls with the velocity v = 4m/s. Its rotational energy is neglected. Its rotational energy is neglected (rotational energy see chapter 7.7).

    a) The rolling friction coefficient is μ_{rr} = 0.02. Calculate the braking force and the braking acceleration.

    b) How far does the ball roll?

4.5 The Slope Down Force and Friction on the Slope

slope, gradient, acclivity	Hang	gravitational force F_g	Gravitationskraft F_G
slope down	den Hang hinunter	downhill-slope force F_h	Hangabtriebskraft F_H
downhill-slope	Gefälle	slope down force, downhill force	Hangabtriebskraft
uphill-slope	Steigung	normal force F_n	Normalkraft F_N
steep slope	Abhang, Steilhang	friction force F_f	Reibungskraft F_R
center of mass	Schwerpunkt	tractive force F_t	Zugkraft F_Z

On a slope, the weight force is divided into two components: The slope down force and the normal force (pressure force).

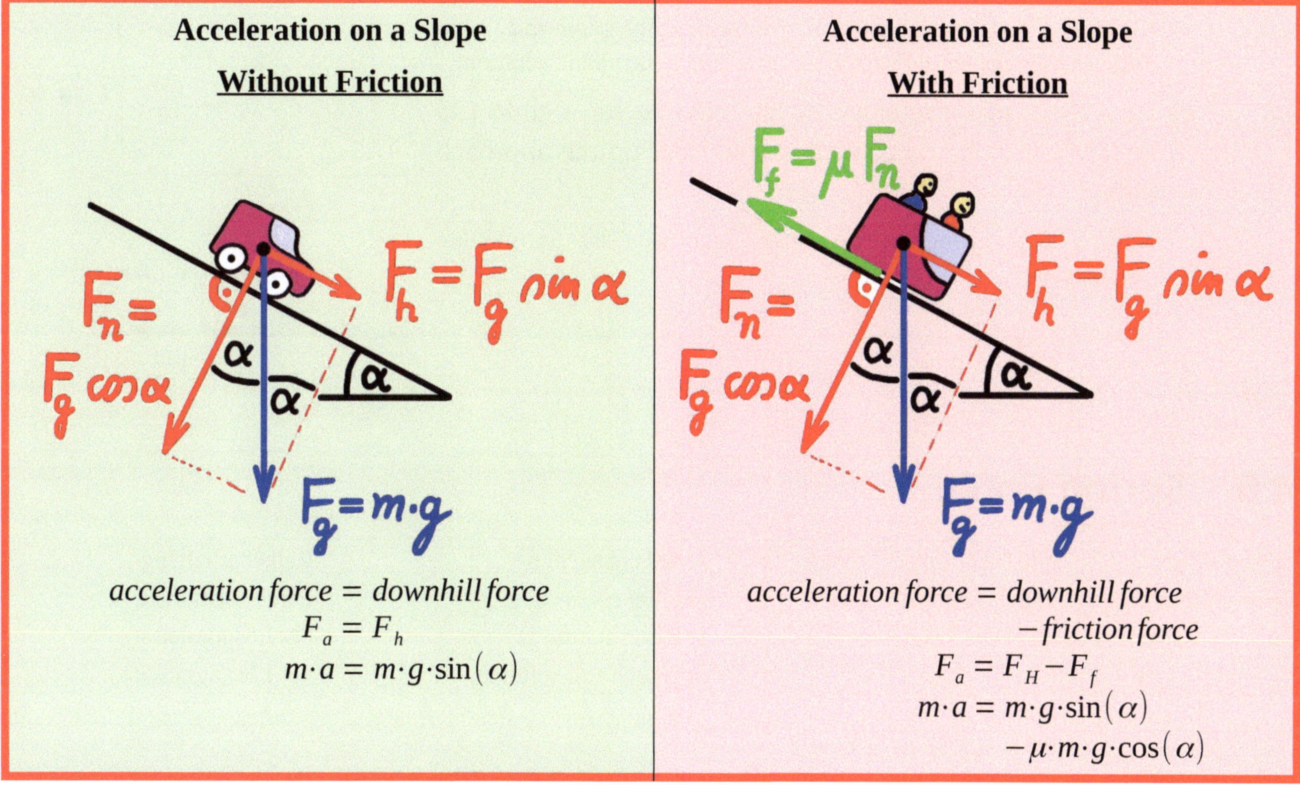

Table 4.3: Friction on a Slope

EXERCISES:

1. **Acceleration on a Slope, without Friction:** on a ski slope, a skier (m = 80kg) is pulled up the slope by a ski lift (gradient 20%). The skier glides almost without friction.

   a) Draw a sketch with all forces.

   b) Calculate the angle of inclination α.

   c) What is the tractive force of the lift when it moves uniformly? (It is assumed that the rope pulls parallel to the slope).

   d) The skier is now pulled up the slope by a rope with a force of 200 N. What is his acceleration now?

   e) After what distance does it reach a speed of 30 km/h?

2. **Acceleration on a Slope, with Friction:** In an amusement park, a slide is to be designed so that the visitors reach a maximum of 30 km/h at the end of the slide. The kinetic friction coefficient fabric-metal is $\mu_k = 0.2$. The slide is 50m long.

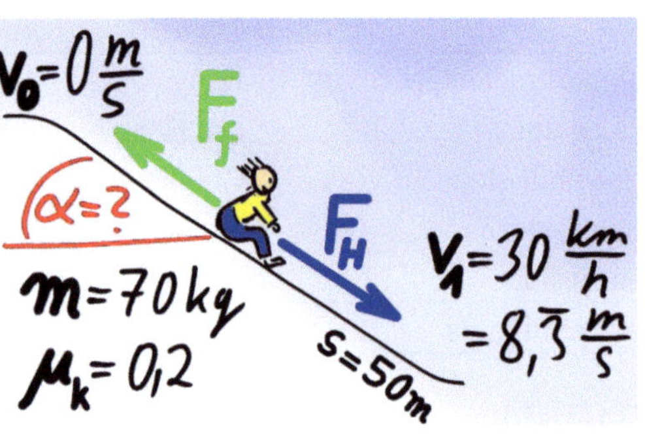

a) Sketch the person sliding down the slide with all the forces acting on it.

b) What is the maximum acceleration allowed to reach only 30 km/h at the end of the slide?

c) How large would the angle have to be at this acceleration? Only set up the approach.

   Note: The complete solution is mathematically somewhat more difficult (trigonometric Pythagoras, substitution, quadratic addition).

d) How large may the angle of the slide be for a child (35 kg)? What is the percentage slope? Set up the approach only.

   Note: For the complete calculation you need the substitution $x = \cos(\alpha)$

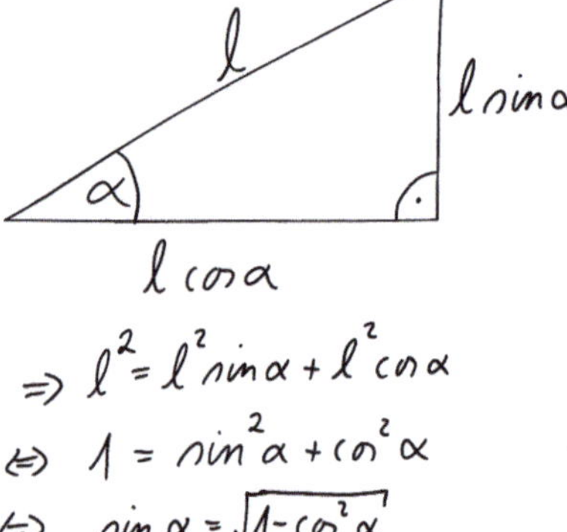

4.6 Spring Force

spring force	die Federkraft	series connection	Reihenschaltung
tension force	die Spannkraft	parallel connection	Parallelschaltung
spring constant	die Federkonstante	solution approach	der Lösungsansatz
		approach to a solution	der Lösungsansatz

The following formulas apply to springs:

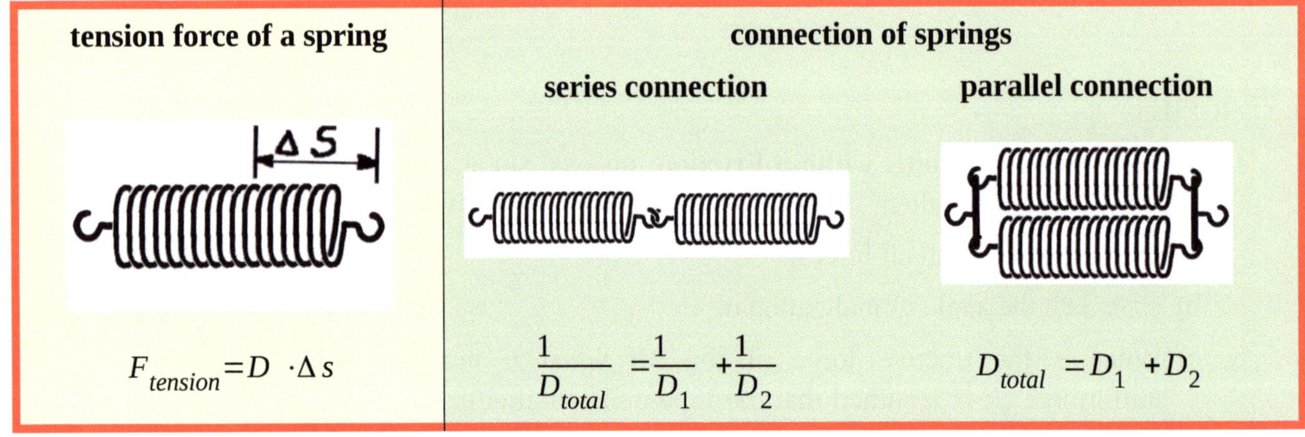

Table 4.4: Spring Forces

In **Volume 1:** "Statics and Hydrostatics" you will find the derivation of the formulas mentioned.

Volume 2: "Kinematics, Dynamics and Energies" will now deal specifically with acceleration and velocities achieved by spring forces.

- ■ **Calculation of Acceleration:**

$$a = \frac{F}{m} = \frac{D \cdot s}{m}$$

- ■ **Calculation of Velocities:**

Force approach:

$$v = \int_{t_1}^{t_2} a(t)\, dt \quad\quad mit\ a(t) = \frac{F(t)}{m}$$

(complicated and impracticable)

Energy approach:

$$E_{tension} = E_{kin}$$

$$\Leftrightarrow \frac{1}{2} \cdot D \cdot s^2 = \frac{1}{2} \cdot m \cdot v^2$$

$$\Leftrightarrow v = \sqrt{\frac{D \cdot s^2}{m}}$$

Energies → see chapter 5.4

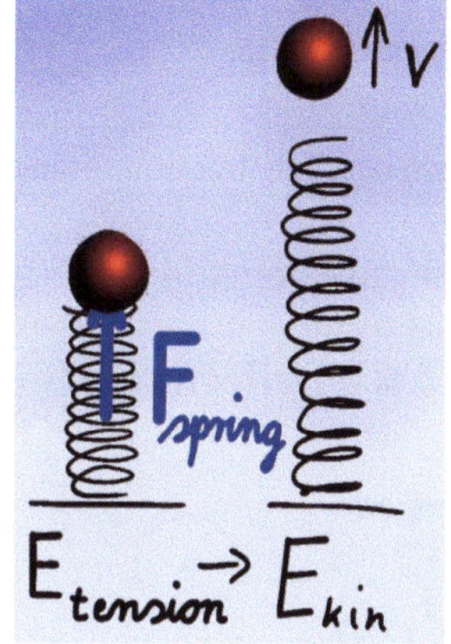

EXERCISES:

1. **Series Connection of Springs:** In a box spring bed, two mattresses lie on top of each other. The upper mattress has a spring constant of D =50 N/cm, the lower mattress has a spring constant of D=46 N/cm. What is the common spring constant D_{total}?

2. **Pea Gun - Acceleration with a Spring:** A spring with the spring constant D = 2 N /cm is tensioned by 4 cm. A pea (ball) of mass 3 g lies on it.

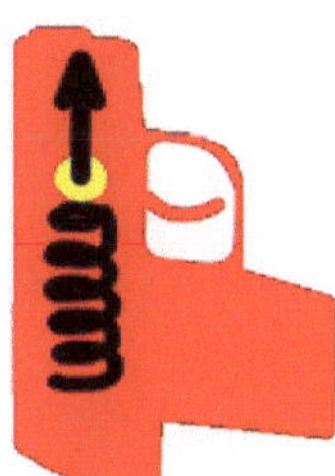

   a) What is the force on the ball at the moment when the spring is released?

   b) What is the acceleration of the pea at this moment? (Assumption: the mass of the spring is almost zero).

   c) What is the acceleration of the ball at the moment when the spring is only tensioned by 3 cm?

4.7   The Rope Force and the Chain-Block

rope force	die Seilkraft	to switch the force round	eine Kraft umlenken
loose pulley	die lose Rolle	to turn the force round	eine Kraft umlenken
fixed pulley	die feste Rolle	to deflect a force	eine Kraft umlenken
chain-block	der Flaschenzug	switching	das Umlenken
transmission ratio	das Übersetzungsverhältnis	inclined plane	die schiefe Ebene
traction	die Zugkraft	ramp	die Rampe
tractive force, traction force	die Zugkraft		

The same tensile force acts at any point along a tensioned rope.

If the rope is guided almost frictionless via deflection pulleys, only the direction of the rope changes, but not its tensile force.

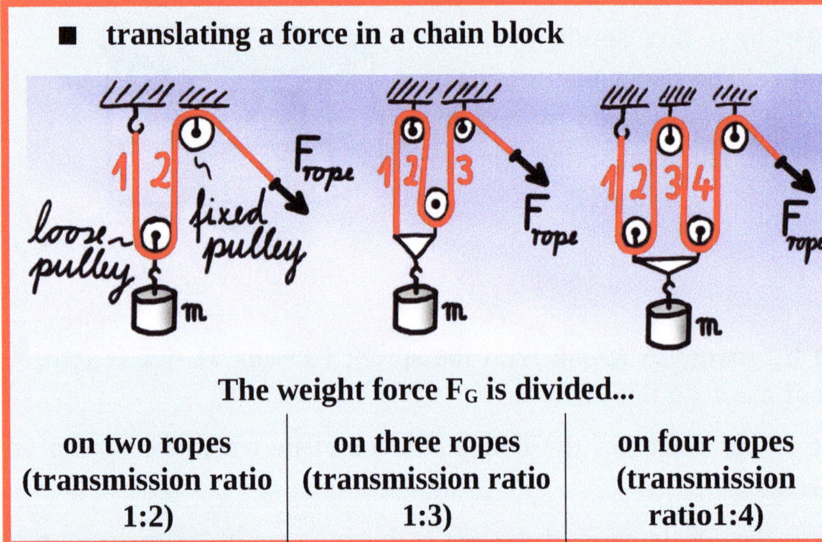

The weight force F_G is divided...

on two ropes (transmission ratio 1:2)	on three ropes (transmission ratio 1:3)	on four ropes (transmission ratio1:4)

A chain block distributes the weight force evenly over several ropes.

The force to be applied becomes smaller, the distance becomes larger.

Table 4.5: Rope Forces

To make the calculation easier, it can help to simplify a drawing:

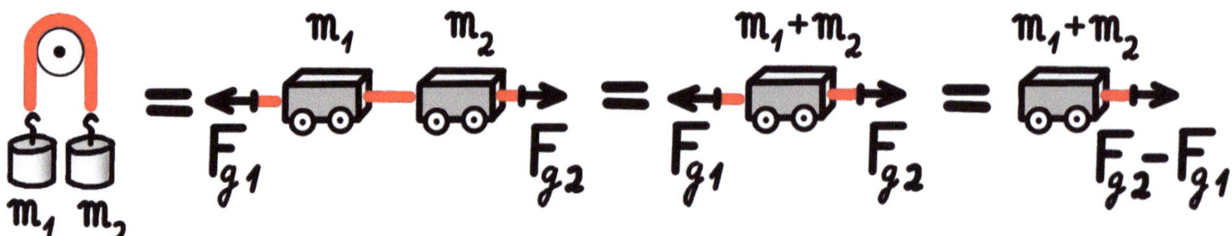

EXERCISES:

1. **Acceleration on the Rope:** What is the acceleration and rope force of the Teddy system shown on the right?

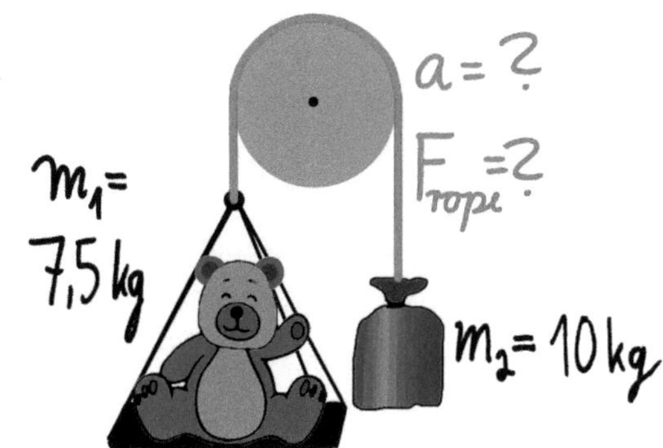

2. **Acceleration on the Inclined Plane:** a toy car (m = 300g) is attached to a traction rope on an inclined plane ($\alpha = 30°$). The traction rope is connected to a weight (m = 200g) via a pulley.

a) First, the car is stationary. Does it accelerate upwards or downwards when you let go of it?

b) What speed does the car reach after traveling 1m?

c) Does the car start moving even if the rolling friction coefficient is $\mu_{rr} = 0.05$? If so, what is the acceleration?

3. **Acceleration in a Chain Block:** The chain block shown on the right is to pull up a bag ($m_1 = 50$ kg).

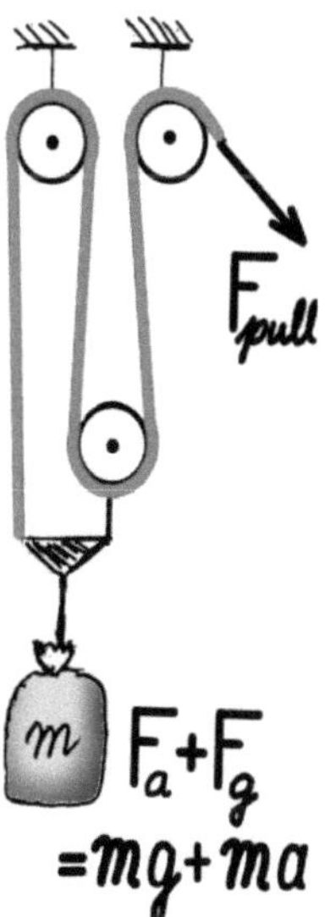

a) With what force must it be pulled?

b) How many meters of rope are pulled if the sack is lifted by 3 m?

Instead of the tractive force F_{pull}, the rope is now pulled by a 40 kg mass m_2.

c) What is the relationship between the accelerations of the masses m_1 and m_2?

d) What is the acceleration of this mass m_2 when it is released? What is the acceleration of the mass m_1 then?

4.8 Exercises

1. **Acceleration Force in a Container Ship:** large container ships can hold more than 20,000 containers with a total mass of up to 200,000 tons and have a length of about 400 meters. The speed is about 40 to 45 km/h, newer ships travel somewhat slower due to fuel prices. The engine power of the largest ships ranges from 50,000 to 100,000 kilowatts. The total mass of the loaded ship is equal to the mass of the displaced water and is therefore also called "displacement".

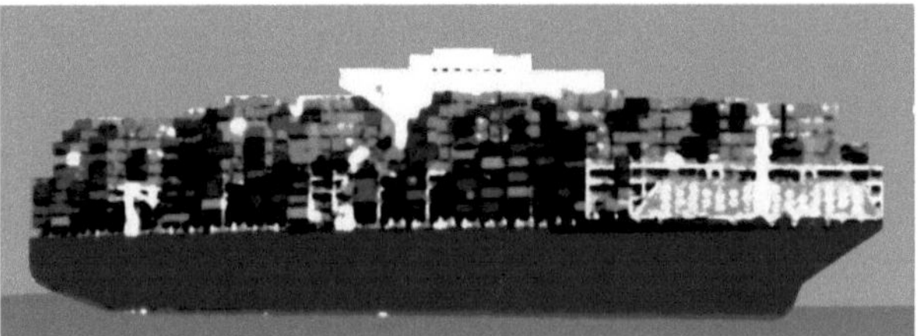

   a) A ship with a displacement of 220,000 tons (= ship mass + container mass) requires 10 kilometers of acceleration travel to reach its maximum speed of 40 km/h. Calculate the acceleration of the ship.

   b) The friction is to be neglected: What force is then used to accelerate the ship?

   c) Calculate the acceleration time in minutes.

   d) To save fuel, the cruising speed is reduced to 30 km/h. How long does the acceleration take now?

2. **Acceleration Force in a Soap Box Race:** Soap boxes made of wood used to be built for packing and transporting soap. Children screwed wheels on them and invented soapbox racing. Today, a "soapbox" is understood to be a vehicle without an engine that only runs on a slope.

   a) The slope of the hill is 20 degrees. Make a sketch with slope down force, normal force and weight force.

   b) A soapbox including the driver weighs 50 kilograms. What is the slope down force and acceleration if friction is neglected?

   c) After what distance has the soapbox accelerated to 20 kilometers per hour?

   d) The coefficient of friction is now $\mu = 0.05$. How large is the friction force?

   e) What distance does the soapbox now need to travel before it reaches 20 kilometers per hour?

5 Energy, Momentum and Efficiency

5.1 Work, Energy and Power

power	Leistung	internal energy	innere Energie
work	Arbeit	kinetic energy	kinetische Energie (Bewegungsenergie)
energy	Energie	potential energy, position energy	potentielle Energie, Lageenergie
efficiency	Wirkungsgrad	thermal energy, heat energy	Wärmeenergie
pulse, impulse, momentum	Impuls	tension energy	Spannenergie

■ **Work and Energy**

work W		energy E	
acceleration work	→	kinetic energy	$W_a = \Delta E_{kin}$
lifting work	→	potential energy	$W_{lift} = \Delta E_{pot}$
friction work	→	thermal energy	$W_f = \Delta E_{th}$
tension work	→	tension energy	$W_{tension} = \Delta E_{tension}$
compression work	→	compression energy (= internal energy)	

Even though the examples show that work and energy are not exactly identical:

In physics, the terms work and energy are usually used as synonyms:
W = E

■ **Most important Energies (Derivation see below)**

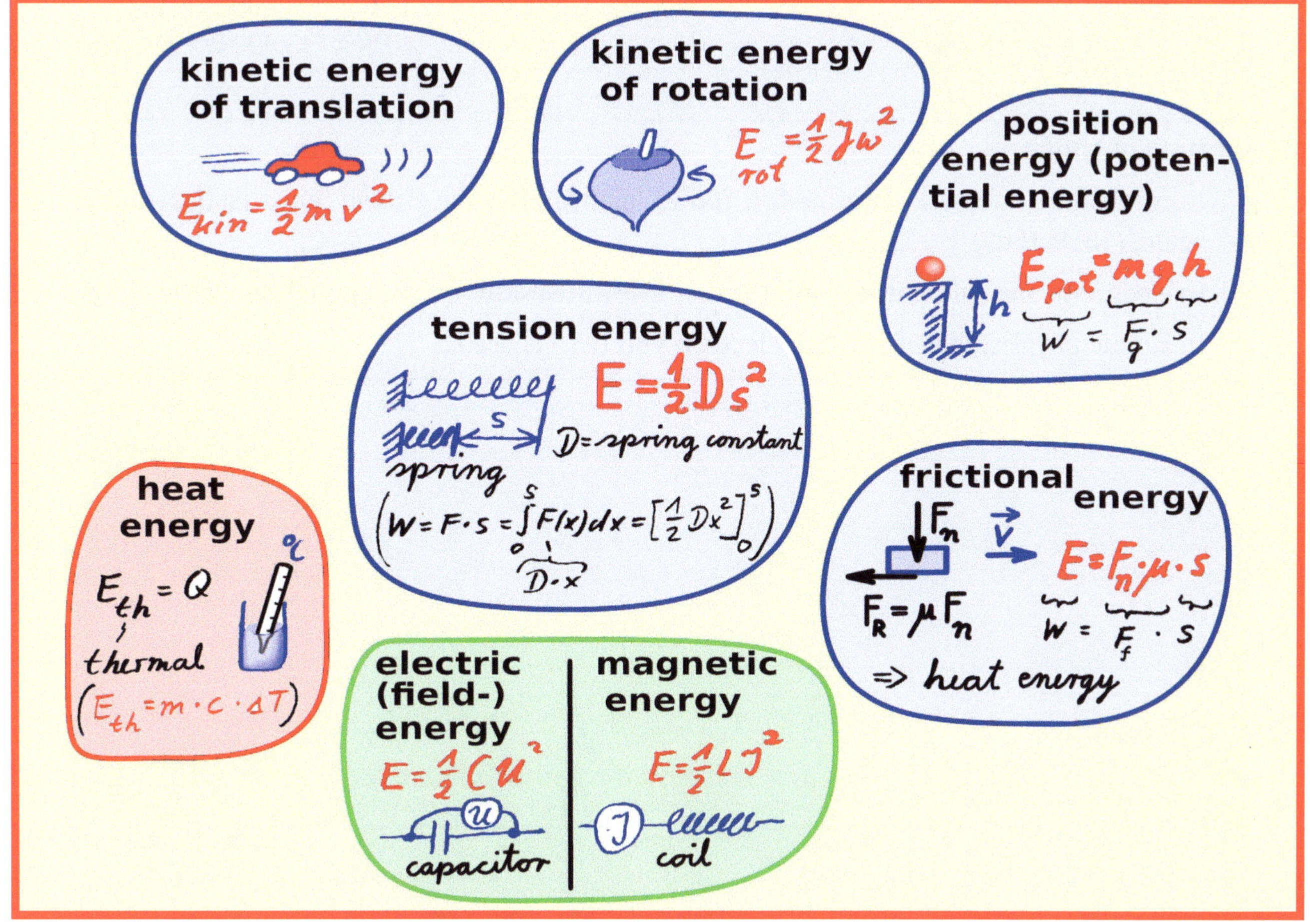

Table 5.1: Most Important Energies

■ Power

<table>
<tr><td>

Power = work per time:

$$\text{Power } P = \frac{W}{t} = \frac{work}{time} = \frac{energy}{time}$$

$$\text{in } \frac{W}{s} = \frac{J}{s} = \frac{Nm}{s}$$

</td></tr>
</table>

Example: The Mercedes 280 SE from 1969

- ...has the **power** of 147 kilowatts.

- ...requires an **energy** of $147\,kW \cdot 10\,s$ to drive 10s at full load

- ...requires on average the **energy** from 13 liters of gasoline per 100 km.

Units	
energy	power
$Nm = J = Ws$	$\dfrac{Nm}{s} = \dfrac{J}{s} = W$
kWh	kW
eV	$\dfrac{eV}{s}$

Conversion of Units:

$$1\,kWh = 3.600.000\,Ws$$
$$1\,PS = 735{,}499\,W\ (veraltet)$$

Application of Units:

- In mechanics, energy is measured in **joules**, named after the British physicist James Prescott Joule, (1818-1889).

- In electricity, the unit **watt-second** (Ws) or **kilowatt-hour** (kWh) is commonly used.

- In atomic physics, the energy unit **electron volt** (eV) is used.

 ©Badelt.de

- **Derivations of the Formulas:**

<table>
<tr><td>For the mechanical energy (work) applies:</td><td>

$$energy\ E = work\ W = force \cdot path$$
$$W = \int_0^s F(x) \cdot dx$$

</td></tr>
</table>

From this follows →

Potential Energy	**Friction Energy**

$$W_{pot} = \int_0^s F_{lift}(x)\,dx \qquad |\ F_{lift} = const$$

$$= \left[F \cdot x \right]_0^s$$

$$= F_{lift} \cdot s \qquad |\ F_{lift} = m \cdot g$$

$$= m \cdot g \cdot h$$

$$W_{fric} = \int_0^s F_{fric}(x)\,dx \qquad |\ F_{fric} = const$$

$$= \left[F \cdot x \right]_0^s$$

$$= F_{fric} \cdot s \qquad |\ F_{fric} = F_n \cdot \mu$$

$$= \mu \cdot F_n \cdot s$$

Kinetic Energy	**Tension Energy**

$$W_{kin} = \int_0^s F_a(x) \cdot dx$$

$$= \int_0^s m \cdot a(x) \cdot dx \qquad \left|\ \frac{dx}{dt} = v(t) \Leftrightarrow dx = v(t) \cdot dt \right.$$

$$= \int_0^{t_1} m \cdot a(t) \cdot v(t) \cdot dt \qquad |\ a(t) = const\,;\, v(t) = at$$

$$= \int_0^{t_1} m \cdot a \cdot a \cdot t \cdot dt$$

$$= \left[\frac{1}{2} m \cdot a^2 \cdot t^2 \right]_0^{t_1}$$

$$= \frac{1}{2} m \cdot a^2 \cdot t^2 \qquad |\ a \cdot t = v$$

$$= \frac{1}{2} \cdot m \cdot v^2$$

$$W_{tension} = \int_0^s F_{tension}(x)\,dx$$

$$= \int_0^s D \cdot x \cdot dx$$

$$= \left[\frac{1}{2} \cdot D \cdot x^2 \right]_0^s$$

$$= \frac{1}{2} \cdot D \cdot s^2$$

EXERCISES

1. **The Kinetic Energy of a Passenger Car:** a passenger car (m=1600 kg) has an engine power of 70 kW. The friction is neglected.

   a) What is the kinetic energy of the car when it accelerates maximally for 6 seconds?

   b) What is the speed of the car then and what is the acceleration a during the acceleration phase?

   c) The car gets a new engine. The engine should be selected so that the car can accelerate from 0 km/h to 100 km/h in only 4s. What power must the new motor have?

   Note: Electric motors are destroyed by heat. Because the heating of the motor is not so fast, electric motors can be extremely overloaded for a short time. This is why even small motors can achieve remarkable accelerations for a short time.

2. **The Position Energy of a Snow Avalanche:** In a mountain range a snow avalanche $(20\,m \cdot 20\,m \cdot 1\,m)$ breaks loose. The snow slab has a mass of 200 tons. The avalanche slides 300 m down into a valley (300 vertical meters).

   a) What was the positional energy of the avalanche at the beginning?

   b) The snow slides almost frictionless. In the valley, the position energy therefore completely becomes kinetic energy ($\rightarrow$ see below: Conservation of energy, chapter 5.4). What is the velocity of the avalanche now?

   c) The avalanche slides against a house. What force does this avalanche exert on the house if it comes to a stop within 3s?

   d) Compare the force from part c) with the weight force of a passenger car (1.5 t).

3. **The Heat Energy in a Kettle:** in a kettle, 1 liter of water (1 kg) should be brought to a boil (100°C) within 2 minutes. The tap water has a temperature of 16 degrees.

   a) What is the total energy absorbed by the water? (See table for heat coefficients).

   b) What is the minimum connected load for this kettle?

   c) Would you recommend connecting it to a standard household socket (voltage: 230V; current max. 16A)? (Electrical power P = $U \cdot I$ = voltage times current, see Volume 4: Electricity).

heat capacities of some substances	
material	heat capacity
iron	$0.439\,\dfrac{kJ}{kg\,K}$
concrete	$0.879\,\dfrac{kJ}{kg\,K}$
water	$4.187\,\dfrac{kJ}{kg\,K}$
ethanol	$2.428\,\dfrac{kJ}{kg\,K}$
machine oil	$1.675\,\dfrac{kJ}{kg\,K}$
air	$1.0054\,\dfrac{kJ}{kg\,K}$

4. **Tension Energy - Derivation of the Formula:** A spring is tensioned with the force $F(x) = D \cdot x$ with $D = 150$ N/cm. Calculate the integral $\int_{0\,m}^{0,08\,m} F(x) \cdot dx$. Sketch the calculated area also in a force-displacement diagram. What is the physical significance of the result?

5. **The Tension Energy of a Spring Shock Absorber:** a truck with a mass of 40 tons distributes its load evenly over six hydraulic shock absorbers (symbol image: small, hydraulic shock absorber).

    Shock absorber principle: In addition to the spring (outside), which cushions unevenness, a piston in an oil-filled cylinder (inside) provides friction. Without friction, the spring would oscillate endlessly. Defective shock absorbers can throw cars off the curve because wild oscillations cause loss of traction.

    a) How large must the spring constant D of a shock absorber be if it is compressed by 20 cm by the weight force?

    [Additional task: Why can't task a) be solved with the energy approach $mgh = 0,5 \cdot D \cdot s^2$? Argue with position energy, stress energy and friction energy. Or: Draw a damped sinusoidal oscillation and mark in this oscillation where the truck is located at the beginning (position energy) and at the end (tension energy)].

    b) What is the total stress energy in the shock absorbers?

    c) What would be the total stress energy in the shock absorbers of this truck if the manufacturer had chosen a spring constant twice as large (note: spring travel changes)?

    d) What physical principle ensures that the tension energy is dissipated after potholes?

6. **Horsepower (obsolete unit):** A body (m = 75 kg) is falling in the gravitational field of the earth (g=9.80665 m/s²). Its velocity is momentarily such that it travels one meter in one second.

    a) What is the acceleration power (m = 75kg; g = 9.80665 m/s²; t = 1s; s = 1m)?

    This acceleration power described above is the definition of "one horsepower". It was formerly used (until 1970 in East Germany and until 1978 in West Germany) to indicate power of engines. In Great Britain and the USA, the original horsepower HP (Horsepower) introduced by James Watt is still used. It differs slightly from horsepower PS in continental Europe because it is traced back to the English units of feet and pounds.

    b) Name the conversion factor between kilowatts and horsepower.

7. **The Frictional Energy of a Disc Brake:** the disc brakes of a truck have a coefficient of sliding friction of $\mu = 0{,}8$. The 7.5 t truck is to be braked from 80 km/h to 0 km/h within 5 s.

a) What is the total energy that the brakes must absorb?

b) How long is the braking distance of the vehicle?

c) What is the braking force for the entire vehicle and what is the braking force at one of the four wheels?

d) The brake shoes are centered on the tire radius. They therefore cover exactly half the braking distance. How great is the braking force there and with what normal force must each of the four brake shoes be pressed against the brake discs?

e) How large is the braking power?

f) What physical principle can be used to generate such a high pressure force?

8. **Power, Fuel Consumption and Efficiency of a Car:** A car has a maximum power of 75 kW. On the daily way to work, however, only 12 kW of power are needed on average.

a) How much power does the car require over the course of a month if the month has 20 working days and the trip to work (there and back) takes 50 minutes each day? Give the solution in kWh and in kJ.

b) The car has an internal combustion engine. Such engines have on average 25% efficiency. How much energy must be supplied to the car (in the form of gasoline) each month?

c) Gasoline has 12 kWh per kg and a density of 0.74 kg per liter. How many liters of gasoline are needed in that month?

d) Alternatively, a car with an electric motor is used. Electric motors have an efficiency of 97%, batteries have an efficiency of 95%. $\left(\rightarrow 0{,}97 \cdot 0{,}95 = 92{,}15\,\% \right)$. How much electrical energy is needed per month?

e) Research prices for electricity and gasoline and compare.

9. **Unit of Power:** A newspaper article states: An average household uses 200 kilowatts of electricity per month. Take a stand on this.

10. **The Thermal Energy in a Disc Brake:** The brakes of a passenger car must be designed to drive downhill on a mountain pass road in the Alps.

Example: In the village of Lofer (600m high) in Austria, a pass road leads up to the Lofer Alm (1400m high). At the top there are alpine huts and a bathing lake, so that tourists as well as suppliers have to use the road every day. A passenger car (1.4 t) drives the road downhill. With the selector lever of the automatic transmission in position "B = engine brake" 40% of the position energy is absorbed by the engine.

a) What is the total energy that the brakes of the passenger car have to absorb?

b) How many meters of altitude per second does the passenger car manage at speed 30 (km/h) and an average gradient (negative slope) of 25%?

c) The brake discs have a total mass of 6kg and a heat capacity of 0.477 kJ / kg K. How warm would the brakes theoretically get if they were not cooled and were 20°C warm beforehand?

d) How far will the car travel (altitude) if the brakes are allowed to heat up by a maximum of 280K to 300°C?

e) [Optional: The brakes are cooled. How far does the car get (difference in altitude) if a constant 1 kW of heat can be dissipated?]

11. **Power and Energy in Mountaineering:** A hiker (95 kg with a backpack) can do about 100 watts as a trained person.

a) Convert the power into the units kW , Ps, kJ/s and J/s. (1 kW = 1.36 hp).

Since 1/1/2010, power in the EU must be expressed in watts or joules/second. The indication of horsepower is only allowed as an addition.

b) What is the maximum number of meters of altitude the climber can climb within three hours? What energy in kWh in kJ and in kcal does this correspond to? (1 kcal = 1 kilocalorie = 4.187 kJ).

c) There are 2245 kJ (or 536 kcal) in a bar of chocolate. How many meters of altitude can the hiker climb with the energy of one bar of chocolate?

d) One kg of body fat has about 7000 kilo-calories (kcal), or about 29,300 kJ. Assume that only body fat is burned: How much body fat has the climber lost if he hikes 2000 meters uphill?

Photo: Path to the „Großes Hundshorn", Lofer (Austria, 2022).

5.2 The Efficiency

work or energy	Arbeit oder Energie	Energy supplied (in) W_{in}	zugeführte Energie W_{zu}
efficiency	Wirkungsgrad	energie delivered (out) W_{out}	abgegebene Energie W_{ab}
incandescent lamp	die Glühlampe	waste heat	Abwärme
incandecence, glow	das Glühen	thermal energy	die Wärmeenergie

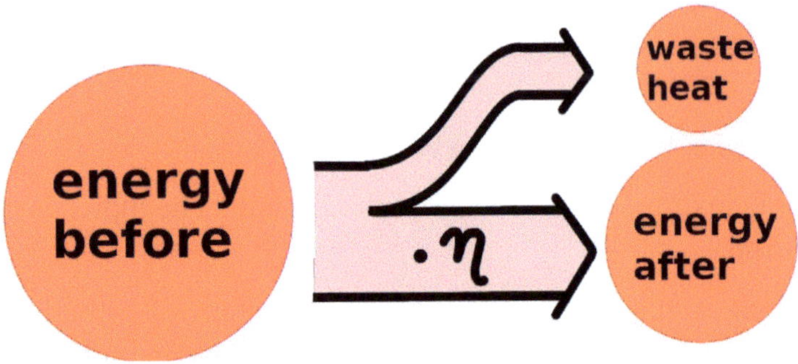

A machine (e.g. an engine, a power plant, a cable car...) always converts a part of its energy into waste heat (= thermal energy).

The part of the used energy is called efficiency η (Eta). η is usually given in percent.

$$energy: \quad E_{in} \cdot \eta = E_{out}$$

$$power: \quad P_{in} \cdot \eta = P_{out}$$

E_{in} = supplied energy (before)
E_{out} = delivered energy (after)

P_{in} = supplied power (before)
P_{out} = delivered power (after)

η = efficiency

■ **Examples for Efficiencies**

An **electric motor** in a car converts 96% of the electrical energy into rotational energy. Only 4% is lost as heat energy. The motor therefore has an efficiency of η = 96%.

Note: Energy never disappears. If a child "loses" his teddy bear, it is still there somewhere, but useless.

A **coal power plant** converts 45 % of the chemical energy into electrical energy. 65 % of the energy is lost as heat. The power plant therefore has an efficiency of η = 45% (as of 2020).

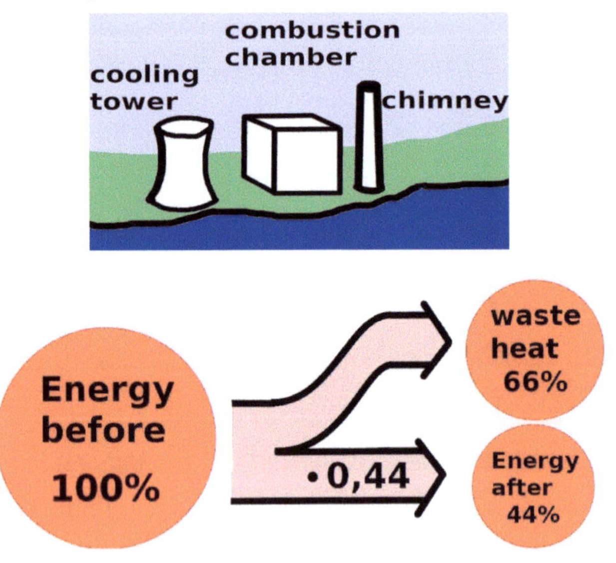

- ■ **Further Efficiencies**

Gasoline engine (at full load)	$\eta = 40\ \%$	Brown coal power plant	$\eta = 40\ \%$
Gasoline engine (on average)	$\eta = 20\ \%$	Hard coal power plant	$\eta = 45\ \%$
Diesel engine (at full load)	$\eta = 55\ \%$	Power plant with district heating	$\eta = 86\ \%$
Diesel engine (on average)	$\eta = 25\ \%$	Electricity grid in Germany	$\eta = 95{,}7\ \%$
Li-ion battery in an electric car	$\eta = 95\ \%$	incandescent lamp	$\eta = 5\ \%$
Electric heater	$\eta = 100\ \%$	LED lamp	$\eta = 30\ ..40\ \%$

Table 5.2: Efficiencies

- ■ **Efficiency Chains (** $\eta_{total} = \eta_1 \cdot \eta_2 \cdot \eta_3 \cdot \ldots$ **)**

Electric car with coal-fired power	Diesel car
(power-plant, electricity-grid, battery, motor)	(refinery, transport-chain, engine)
$45\% \cdot 95{,}7\% \cdot 95\% \cdot 96\% = 39{,}3\%$	$95\% \cdot 90\% \cdot 25\% \ = \ 21\%$

Table 5.3: Efficiency Chains

- ■ **Sources (as of 2018)**

http://www.umweltbundesamt.de/

https://de.wikipedia.org/wiki/Elektroauto#Verbrauch_und_Wirkungsgrad

EXERCISES

1. **Calculate the Efficiency**

   a) 70% of the power consumption becomes unused heat.

   b) A mixer for cake batter consumes 50W of power. Its useful power is 32W.

   c) A charger emits 5W to the battery, generating 1W of heat that is released to the environment. The battery also gets warm while charging and it gives off 0.7 W as heat to the environment.

      - What is the efficiency of the charger?

      - What is the efficiency of the battery during charging?

2. **Efficiency of a Hydroelectric Power Plant:** The city of Las Vegas in the USA has the highest electricity consumption per capita in the world. Advertising, casinos and nightclubs mean that twice as much electricity is consumed at night as during the day. Some of the electricity comes from the hydroelectric power plant at Hoover Dam. Here, at a head of 180m, up to 1470 m³ of water per second flows through the turbines (water: ρ = 1 kg per liter).

   a) What is the output of the hydroelectric power plant if an efficiency of η = 82% is assumed? (Indicate in kJ/s and in kW).

   b) The dam generates 4 billion kilowatt-hours of electricity per year. What is the average output (average power) of the power plant?

3. **Efficiency Chain:** An electric car draws its energy from a coal-fired power plant. The following efficiencies are known: coal-fired power plant: 47%, power grid: 95%, Li-ion battery: 95%, electric motor 97%.

   a) What is the total efficiency of the electric car from coal-fired power to the wheel?

   A diesel vehicle gets its energy from diesel fuel. The following efficiencies are known: refinery: 95%, transport chain: 90%, diesel engine at rated speed: 45%, diesel engine at usual speed in traffic: 25%.

   b) What is the total efficiency of the diesel car from refinery to wheel in usual traffic?

   c) Additional task: By far the worst efficiency is achieved by vehicles (passenger cars) that run on hydrogen (fuel cells), partly because hydrogen requires energy to be pressed into the tank or cooled. Nevertheless, the technology is being pursued by Toyota and Mercedes. Name at least two advantages of hydrogen fuel cell technology.

© *Badett.de*

5.3 The Efficiency of Heat Pumps and Condensing Boiler Technology

condensing (boiler) technology	Brennwerttechnik	favorable, unfavorable	günstig, ungünstig
higher heating value HHV	oberer Heizwert H_O (=Brennwert)	hydrocarbon	Kohlenwasserstoff
(gross energy, upper caloric value)	oberer Heizwert H_U (=Brennwert)	water vapor	Wasserdampf
lower heating value LHV	unterer Heizwert (old: Heizwert)	to condense, to precipiate	kondensieren
(lower caloric value LCV)	unterer Heizwert (old: Heizwert)	heat pump	Wärmepumpe
		annual performance factor	Jahresarbeitszahl

■ The Condensing Technology

Hydrocarbons burn to form carbon dioxide and water. Therefore, when gas is burned, water vapor or water is always produced.

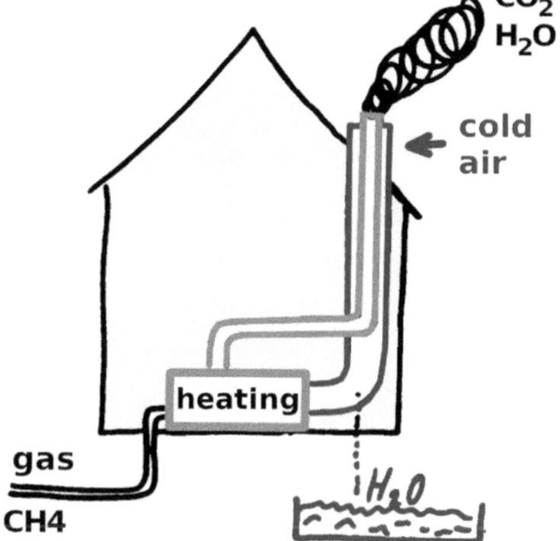

- Unfavorable case: If water vapor is produced, we speak of the **lower calorific value**. The water vapor takes a lot of heat energy with it.

- Favorable case: If water is produced, we speak of the **upper calorific value**: The water contains less energy (calorific value = energy per cubic meter of gas).

By condensing the water vapor in the chimney, it is possible to utilize the upper heating value. However, if the heat gained is then incorrectly related to the lower calorific value, this results in efficiencies of around 104 %. This "error" is common in heating system construction.

Notes:

- Condensing technology is not worthwhile with fuels that contain little hydrogen, e.g. hard coal (LHV /HHV = 0.958).

- Condensing technology is not worthwhile if the heating system runs out of the condensing range. This happens, for example, at very high flow temperatures or when many heaters are connected to the chimney and consequently the very warm exhaust gas no longer condenses.

■ **The Heat Pump**

Heat pumps, for example, produce a heat output of 9 kW with 3 kW of electrical power. The missing 6 kW are taken from the geothermal heat:

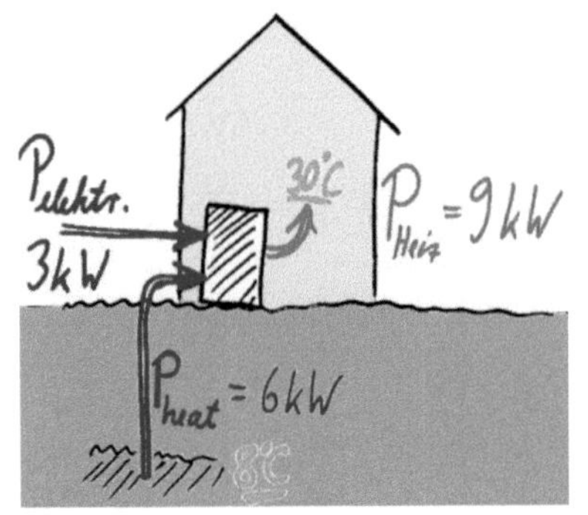

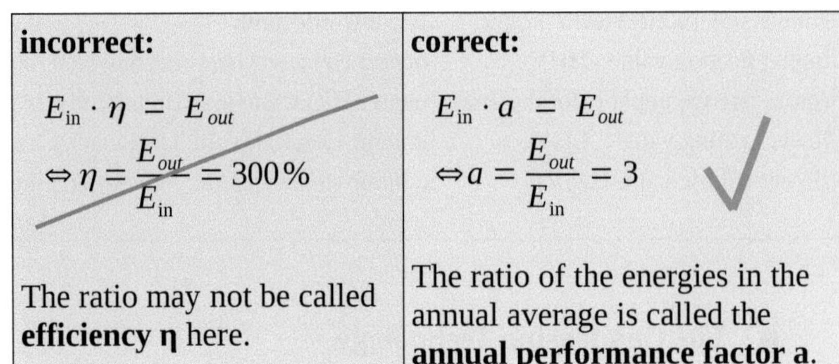

incorrect:	correct:
$E_{in} \cdot \eta = E_{out}$ $\Leftrightarrow \eta = \dfrac{E_{out}}{E_{in}} = 300\,\%$	$E_{in} \cdot a = E_{out}$ $\Leftrightarrow a = \dfrac{E_{out}}{E_{in}} = 3$ ✓
The ratio may not be called **efficiency η** here.	The ratio of the energies in the annual average is called the **annual performance factor a.**

It is important to use the correct terms: "annual performance factor 3" instead of "efficiency 300%". Modern heat pumps achieve an annual performance factor >4 (as of 2022). This also depends mainly on the temperature difference Taußen (deep drilling) and Tinnen (low-temperature surface radiators).

EXERCISES

1. **Heat Pump**

   a) Explain the term "annual performance factor".

   b) While gas heating and oil heating make do with conventional radiators, the heat pump requires underfloor heating or wall heating. State the reason for this.

   c) What is the point of deep drilling for heat pumps?

2. **Condensing Technology:** natural gas consists to a large extent of methane (CH4). Methane can also be produced artificially from carbon dioxide and hydrogen. For this purpose, green hydrogen can be produced from water and regeneratively generated electricity using electrolysis.

   The upper heating value (= calorific value) for methane is 39.82 MJ/m³ at 0 °C.

   The lower heating value for methane is 35.88 MJ/kg at 0 °C.

   a) What percentage more energy can be obtained if the water vapor is condensed after methane combustion?

   b) A household with methane condensing heating requires 1,000 kWh of heating energy per year. Calculate how much of this energy comes from condensing water in the chimney (in the ideal case).

   c) In a very cold winter, the heating is turned up: flow temperature and flue gas temperature increase. What problems arise for condensing technology?

5.4 The Law of Conservation of Energy

law of conservation of energy	der Energieerhaltungssatz	internal energy	innere Energie (Wärme, Druck...)
inclined plane	die schiefe Ebene	potential energy	potentielle Energie (Lageenergie)
angle of inclination	der Neigungswinkel	kinetic energy	kinetische Energie
		frictional heat	Reibungswärme

Law of conservation of energy: The sum of all energies always remains constant.

$$\text{sum of all energies before} \ = \ \text{sum of all energies after}$$

$$\Leftrightarrow \qquad \sum E_{i,before} \ = \ \sum E_{i,after}$$

■ **Example:**

A body slides down an inclined plane. Its potential energy changes into frictional heat and kinetic energy:

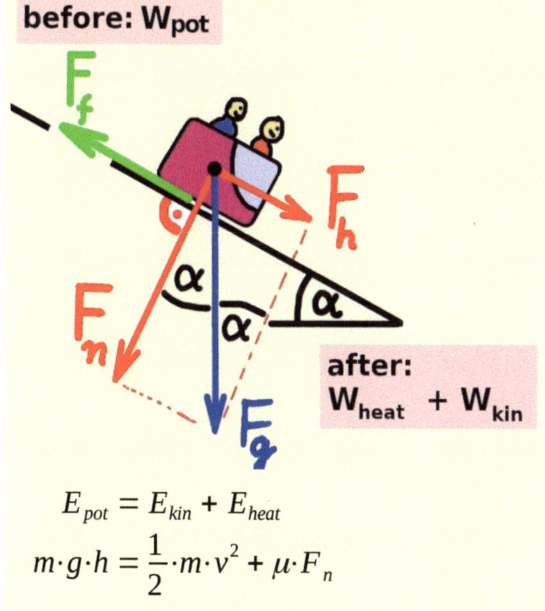

$$E_{pot} = E_{kin} + E_{heat}$$
$$m \cdot g \cdot h = \frac{1}{2} \cdot m \cdot v^2 + \mu \cdot F_n$$

■ **Common Mistakes:**

Energies often hide where you would not suspect them. Examples:

- Pirouette of a figure skater: The rotating figure skater tightens his arms and becomes faster.

  Wrong: Rotational energy before = rotational energy after.

  Correct: Tightening the arms is additional energy due to muscle work.

- Collision of two cars:

  Wrong: Kinetic energies before = Kinetic energies after.

  Correct: The deformation and heating of the sheet metal swallows energy (= internal energy).

- Electrical energy can be stored between two metal plates. (→ capacitor). Now the plates are to be pulled apart.

  Wrong: Electrical energy before = electrical energy after.

  Correct: Pulling the attracting plates apart supplies energy to the system.

EXERCISES:

1. **A Block of Wood slides (frictionlessly) on an Inclined Plane.** The plane has a length of 40 cm with an angle of inclination of 50°. The mass of the block is 0.3 kg.

   a) Determine the velocity reached at the end using the force approach.

      - What is the magnitude of the downslope force?
      - What is the magnitude of the acceleration?
      - What is the velocity reached?

   b) Determine the velocity reached at the end using the law of conservation of energy.

   Let the coefficient of friction now be $\mu = 0.2$.

   c) Determine again the velocity reached at the end using the force theorem.

   d) Find again the velocity reached at the end using the law of conservation of energy.

2. **Heat Capacity and Frictional Heat:** The block from task 1 now slides down from the height $h = 0.306$ m and reaches a velocity of $v = 1$ m/s. Calculate with the law of conservation of energy: By how much does the block heat up if its heat capacity is 0.47 Wh /kgK? To simplify, assume that all the frictional heat remains in the block.

 ©Badelt.de

5.5 Momentum and Force Impact

pulse, impulse, momentum	der Impuls	ball	die Kugel, der Ball
force impact	der Kraftstoß	bullet	die Kugel, das Geschoss
		sphere	die Kugel, das Himmelsgewölbe
		bowl	die Kugel, die Schale

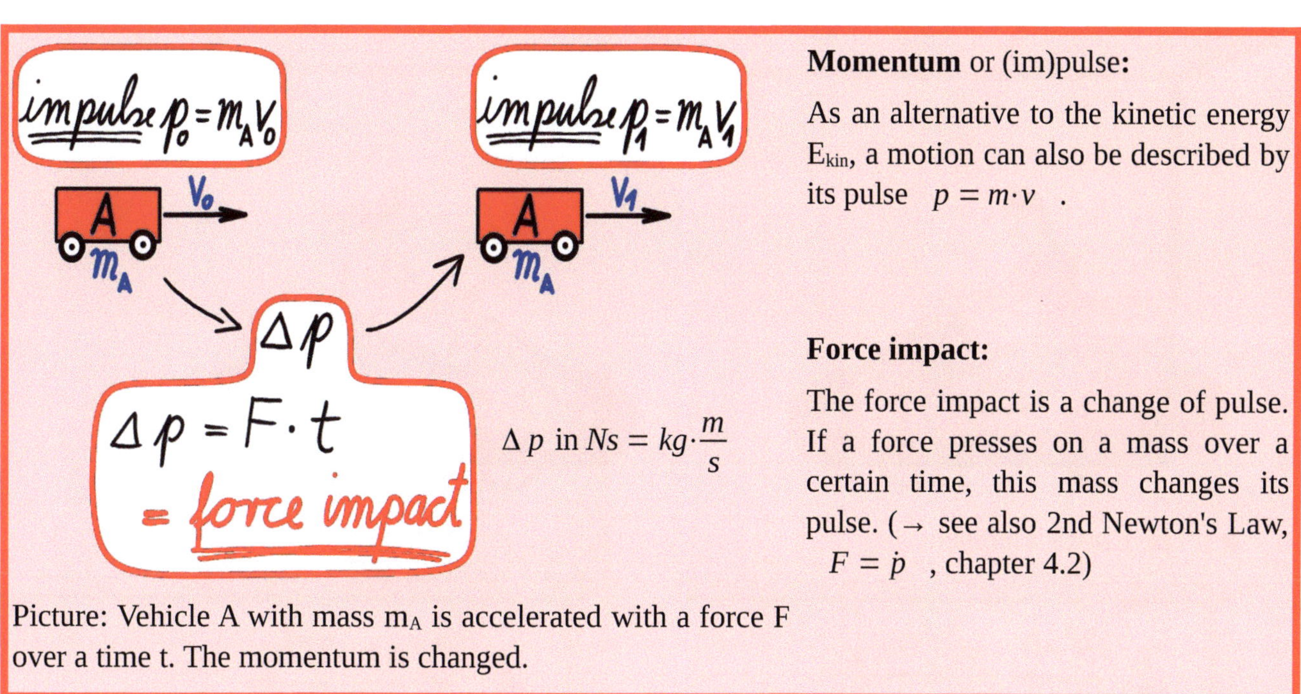

Momentum or (im)pulse:

As an alternative to the kinetic energy E_{kin}, a motion can also be described by its pulse $p = m \cdot v$.

Force impact:

The force impact is a change of pulse. If a force presses on a mass over a certain time, this mass changes its pulse. ($\rightarrow$ see also 2nd Newton's Law, $F = \dot{p}$, chapter 4.2)

Picture: Vehicle A with mass m_A is accelerated with a force F over a time t. The momentum is changed.

Table 5.4: Momentum and Force Impact

EXERCISES

1. **Force Impact on a Passenger Car:** A motionless car (1500 kg) is pushed with the force 600 N for 12 seconds.

    a) What is its momentum now?

    b) What is its velocity?

    c) What would be the speed of a car with half the weight after this force impact?

    d) The force impact is given in Ns. The momentum is given in. How many correspond to one?

2. **Force Impact on a Ball:** A ball weighing 500 g has an impulse of 30.

    a) A force impact of 5 Ns acts in the direction of motion. How fast is the ball now? (the rotation is neglected).

    b) How fast would the ball be if the force impact acted against the direction of motion?

5.6 The Law of Conservation of Momentum

law of conservation of momentum	der Impulserhaltungssatz	elastic collision (elastic impact)	der elastische Stoß
collision, shock pulse, impact	der Stoß	inelastic collision (inelastic impact)	der unelastische Stoß
bumper car	der Autoscooter	inelastic collision (inelastic impact)	der plastische Stoß

■ **The Collision or Impact describes the Collision of two Bodies** (here: bumper cars).

■ **The Elastic Collision**

After an elastic impact, the vehicles are separated again.

■ **The Inelastic Collision**

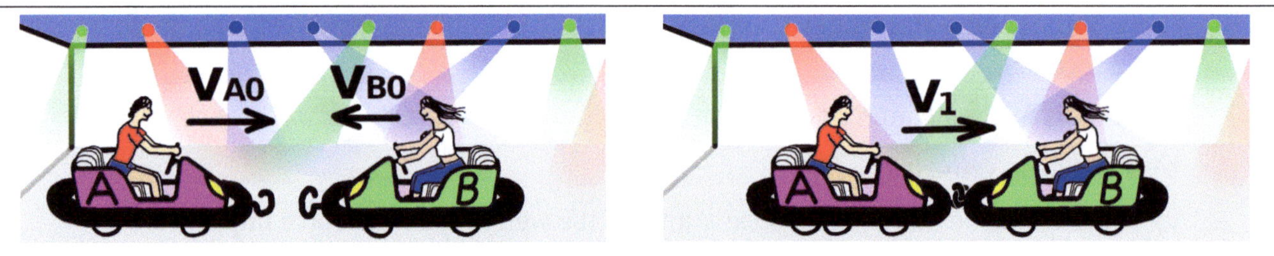

After an inelastic impact, the vehicles hang together.
The coupling becomes warm and absorbs part of the kinetic energy.

The energy absorbed by the vehicle is called "internal energy". It can be heat energy, deformation
energy or stress energy/pressure energy.

- **Variant 1: Representation of the formulas with v_0 before the impact and v_1 after the impact. (You only need one of the two variants).**

<table>
<tr><td colspan="2" align="center">Elastic Collision:</td><td colspan="2">Law of conservation of energy</td></tr>
</table>

Elastic Collision:	Law of conservation of energy
 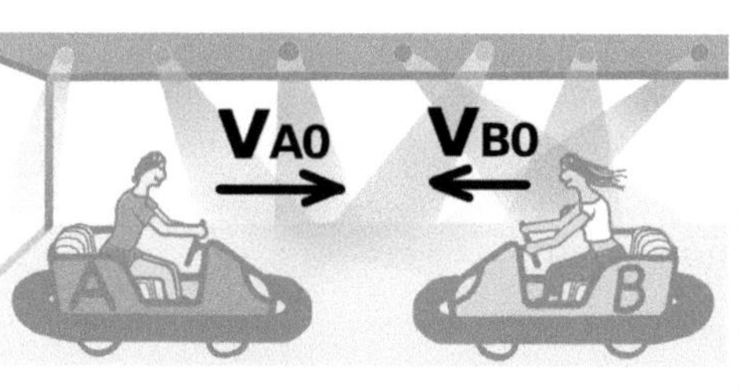 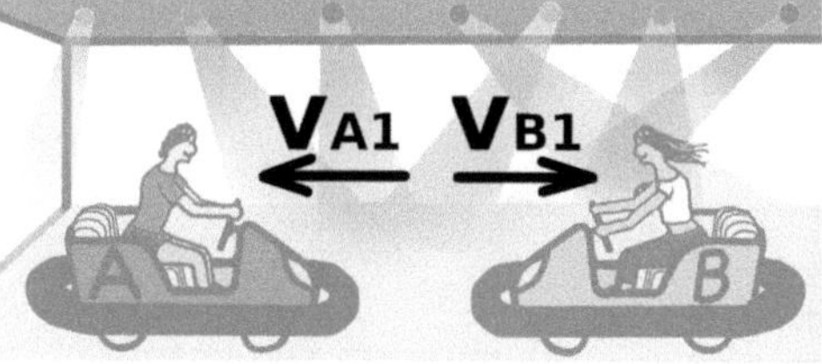	$$E_{kin\,A0} + E_{kin\,B0} = E_{kin\,A1} + E_{kin\,B1}$$ $$\frac{1}{2}\cdot m_A\cdot v_{A0}^2 + \frac{1}{2}\cdot m_B\cdot v_{B0}^2 = \frac{1}{2}\cdot m_A\cdot v_{A1}^2 + \frac{1}{2}\cdot m_B\cdot v_{B1}^2$$ **Law of conservation of momentum** $$p_{A0} + p_{B0} = p_{A1} + p_{B1}$$ $$m_A\cdot v_{A0} + m_B\cdot v_{B0} = m_A\cdot v_{A1} + m_B\cdot v_{B1}$$ **The combination of both formulas leads to:** $$v_{A1} = \frac{(m_A - m_B)\cdot v_{A0} + 2\,m_B v_{B0}}{m_A + m_B}$$ $$v_{B1} = \frac{(m_B - m_A)\cdot v_{B0} + 2\,m_A v_{A0}}{m_A + m_B}$$ * Derivation see below

Table 5.5: Elastic Collision

Inelastic Collision:	Law of conservation of energy
 	$$E_{kin\,A0} + E_{kin\,B0} = E_{kin\,1} + E_{innere}$$ $$\frac{1}{2}\cdot m_A\cdot v_{A0}^2 + \frac{1}{2}\cdot m_B\cdot v_{B0}^2 = \frac{1}{2}\cdot(m_A + m_B)\cdot v_1^2 + E_{innere}$$ **Law of conservation of momentum** $$p_{A0} + p_{B0} = p_1$$ $$m_A\cdot v_{A0} + m_B\cdot v_{B0} = (m_A + m_B)\cdot v_1$$ $$\Leftrightarrow \quad v_1 = \frac{m_A\cdot v_{A0} + m_B\cdot v_{B0}}{m_A + m_B}$$

Table 5.6: Inelastic Collision

- **Variant 2: Representation of the formulas with v before the impact and u after the impact. (You only need one of the two variants).**

Elastic Collision:	Law of conservation of energy
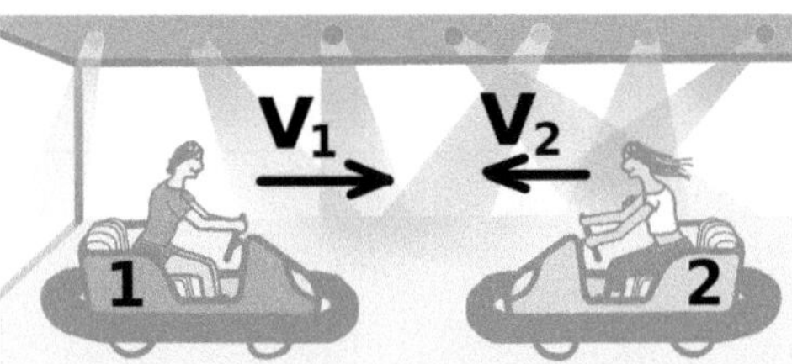 	$$E_{kin1} + E_{kin2} = E_{kin1}{}' + E_{kin2}{}'$$ $$\frac{1}{2}\cdot m_1\cdot v_1^2 + \frac{1}{2}\cdot m_2\cdot v_2^2 = \frac{1}{2}\cdot m_1\cdot u_1^2 + \frac{1}{2}\cdot m_2\cdot u_2^2$$

Law of conservation of momentum

$$p_1 + p_2 = p_1{}' + p_2{}'$$
$$m_1\cdot v_1 + m_2\cdot v_2 = m_1\cdot u_1 + m_2\cdot u_2$$

The combination of both formulas leads to:

$$u_1 = \frac{(m_1 - m_2)\cdot v_1 + 2\,m_2 v_2}{m_1 + m_2}$$

$$u_2 = \frac{(m_2 - m_1)\cdot v_2 + 2\,m_1 v_1}{m_1 + m_2} \qquad \text{* Derivation see below}$$

Inelastic Collision:	Law of conservation of energy
 	$$E_{kin1} + E_{kin2} = E_{kin}{}' + E_{innere}$$ $$\frac{1}{2}\cdot m_1\cdot v_1^2 + \frac{1}{2}\cdot m_2\cdot v_2^2 = \frac{1}{2}\cdot(m_1+m_2)\cdot u^2 + E_{innere}$$

Law of conservation of momentum

$$p_1 + p_2 = p{}'$$
$$m_1\cdot v_1 + m_2\cdot v_2 = (m_1 + m_2)\cdot u$$

$$u = \frac{m_1\cdot v_1 + m_2\cdot v_2}{m_1 + m_2}$$

- **Derivation of the Formulas**

$$I \qquad \frac{1}{2}\cdot m_A\cdot v_{A0}^2 + \frac{1}{2}\cdot m_B\cdot v_{B0}^2 = \frac{1}{2}\cdot m_A\cdot v_{A1}^2 + \frac{1}{2}\cdot m_B\cdot v_{B1}^2 \qquad \left|\cdot\frac{2}{m_A}\right.$$

$$\Leftrightarrow \quad v_{A0}^2 + \frac{m_B}{m_A}\cdot\left(v_{B0}^2 - v_{B1}^2\right) = v_{A1}^2$$

$$II \qquad m_A\cdot v_{A0} + m_B\cdot v_{B0} = m_A\cdot v_{A1} + m_B\cdot v_{B1} \qquad \left|-m_B\cdot v_{B1}\right.$$

$$\Leftrightarrow \qquad v_{A0} + \frac{m_B}{m_A}\cdot\left(v_{B0}-v_{B1}\right) = v_{A1} \qquad \left|\,(...)^2\right.$$

$$\Leftrightarrow \quad v_{A0}^2 + 2\cdot v_{A0}\cdot\frac{m_B}{m_A}\cdot\left(v_{B0}-v_{B1}\right) + \left(\frac{m_B}{m_A}\right)^2\cdot\left(v_{B0}-v_{B1}\right)^2 = v_{A1}^2$$

Equating the formulas: $\qquad v_{A1}^2 = v_{A1}^2$

$$\Leftrightarrow \quad v_{A0}^2 + \frac{m_B}{m_A}\cdot\left(v_{B0}^2 - v_{B1}^2\right) = v_{A0}^2 + 2\cdot v_{A0}\cdot\frac{m_B}{m_A}\cdot\left(v_{B0}-v_{B1}\right) + \left(\frac{m_B}{m_A}\right)^2\cdot\left(v_{B0}-v_{B1}\right)^2 \quad \left|-v_{A0}^2\;\right|\cdot\frac{m_A}{m_B}$$

$$\Leftrightarrow \qquad \left(v_{B0}^2 - v_{B1}^2\right) = 2\cdot v_{A0}\cdot\left(v_{B0}-v_{B1}\right) + \left(\frac{m_B}{m_A}\right)\cdot\left(v_{B0}-v_{B1}\right)^2 \qquad |\,3.\ \text{binom. Formula}$$

$$\Leftrightarrow \quad \left(v_{B0} - v_{B1}\right)\cdot\left(v_{B0} + v_{B1}\right) = 2\cdot v_{A0}\cdot\left(v_{B0}-v_{B1}\right) + \left(\frac{m_B}{m_A}\right)\cdot\left(v_{B0}-v_{B1}\right)^2 \qquad \left|\cdot\frac{1}{v_{B0}-v_{B1}}\right.$$

$$\Leftrightarrow \qquad \left(v_{B0} + v_{B1}\right) = 2\cdot v_{A0}\cdot + \left(\frac{m_B}{m_A}\right)\cdot\left(v_{B0}-v_{B1}\right) \qquad \left|\cdot m_A\right.$$

$$\Leftrightarrow \quad m_A\cdot v_{B0} - 2\,m_A v_{A0} - m_B v_{B0} = m_A v_{B1} - m_B v_{B1} \qquad \left|\cdot\frac{1}{m_A+m_B}\right.$$

$$\Leftrightarrow \quad \frac{\left(m_A - m_B\right)\cdot v_{B0} - 2\,m_A v_{A0}}{m_A + m_B} = -v_{B1} \qquad \left|\cdot\frac{1}{m_A+m_B}\right.$$

$$\Leftrightarrow \quad \frac{\left(m_B - m_A\right)\cdot v_{B0} + 2\,m_A v_{A0}}{m_A + m_B} = v_{B1}$$

EXERCISES:

1. **Elastic Collision:** A toy car (m1= 300g, v1= 5 m/s) collides completely elastically with an oncoming car (m2= 200g, v2= -3 m/s). What are the velocities of the cars after the collision?

2. **Rifle Bullet hits Sandbag:** A rifle bullet (v = 400 m/s; m = 50g) hits a stationary sandbag (m=8 kg) hanging on a rope. The bullet gets stuck in the bag.

   a) Is this an elastic or an inelastic impact?

   b) What speed do the sandbag and the ball reach together?

   c) By how many degrees is the sandbag deflected when the rope is three meters long?

3. **Elastic Impact, Inelastic Impact and Thermal Energy:** At a marshalling yard, a 15-ton rail car rolls down a sloping track. The inclination is 2 degrees to the horizontal, the track is 20 meters long. Friction is negligible.

   a) What kinetic energy and what speed does the wagon reach when it reaches the bottom? (before the wagon stands still: $v_0 = 0$).

   b) At the bottom, the wagon meets a stationary second wagon weighing 25t. There is no coupling. What are the velocities of both wagons after the collision?

   c) What would be the speed if both wagons were coupled at the impact?

   d) During a inelastic impact (coupling), part of the kinetic energy is always lost as heat. What thermal energy would have to be released during this coupling process?

   e) If the internal energy is only half as great (and the speed after the impact remains the same), what is the speed of the wagon before the inelastic impact?

   f) What is the velocity of the wagon after the impact if it hits a rigid concrete wall completely inelastically?

4. **Impulse ($p = m \cdot v$) at Rocket Engine:** In space usual aircraft engines do not work. A spaceship can be accelerated there only with the help of gravitation (it follows a planet which attracts it), or with the help of the law of conservation of momentum (it uses rocket engines or solar sails).

   a) **Function of the Rocket Engine:** Julius sits in a boat, has lost his oars and throws stones backwards to accelerate the boat. In this comparison Julius is the astronaut, the boat is his spaceship and the stones are his fuel. The boat with astronaut and stones weighs 90 kg. One stone weighs 0.5 kg. What speed does the (initially stationary) boat reach when a stone is thrown backwards at 10m/s?

   b) **Acceleration of a Satellite (reference system = initial satellite position):** A satellite (m = 2.5 t) has to correct its orbit. For this purpose a rocket engine is ignited. This ejects 300g of propellant at a velocity of 1500 m/s. To what speed is the satellite accelerated? (Let the reference system be the location of the satellite before the course correction). Calculate with the law of conservation of momentum.

   c) **Acceleration of a Satellite (reference frame = earth):** The satellite is far out in space moving away from the Earth at a speed of 1200 m/s. The course correction is to slow it down. What speed does it reach after the correction?

5.7 Force Impact for Jet Engine and Rocket Engine

pulse principle	das Impulsprinzip	rocket engine	Raketentriebwerk
gravity principle	das Gravitationsprinzip	jet engine	Strahltriebwerk
rocket equation	die Raketengleichung	plasma engine	Plasmatriebwerk
		solar sail	Sonnensegel

■ The Acceleration of a Rocket in Space

Pulse Principle (carried fuel):

- Rocket engine: Matter is pushed backwards with great speed, with the help of a chemical reaction. The sum of all impulses remains zero (see right).

- Plasma engine (magnetic field oscillation engine): Matter is thrust backward at great velocity, with the aid of a magent field. Significantly less fuel required and less power.
  For very long distance flights (Mars). Energy from solar panel.

Pulse Principle (without fuel):

- Solar sails: The momentum of photons emitted by the sun is used. Lasers from Earth can also theoretically drive a solar sail.

- Research is still underway to exploit the photon momentum of light particles generated on board, which would of course mean that there would be no need for fuel on board. Light particles can be generated from pure energy.

Gravity Principle:

- Of course, nowadays mainly the gravity of moving celestial bodies is used for acceleration.

■ The Calculation of the Acceleration Force of Rockets and Jet Engines

Power through fuel output:

$$Force\ Impact = Momentum\,(=impulse)$$

$$\Leftrightarrow \quad F \cdot t = m_{fuel} \cdot v_{fuel} \quad | \quad m_{fuel} \ll m_{rocket} \rightarrow v_{fuel} = v_{relative}$$

$$\Leftrightarrow \quad F \cdot t = m_{fuel} \cdot v_{rel} \quad | \quad \dot{m}_{fuel} = \dot{m}_{rocket} = \dot{m}$$

$$\Leftrightarrow \quad F = \dot{m} \cdot v_{rel}$$

With $\quad F = m \cdot a \quad$ this results in the **rocket equation:** $\quad v_{end} = v_{rel} \cdot \ln\left(m_2 - m_1\right)$

$$(\rightarrow \text{full derivation see Volume 6, Mathematics in Physical Applications}).$$

Table 5.7: Rocket Equation

EXERCISES

1. **Force Impact ($F \cdot t = m \cdot v$) in the Jet Engine:** The jet engine of an aircraft is to generate a thrust force of 190 kN (=force impact). For this purpose, the air is pushed backward at a speed of 210 m/s (relative to the aircraft). (Density of air = 0.00129 g/cm³).

    a) How many kg of air must be passed through the engine per second for this purpose?

    b) How many m³/s is that?

2. **Force Impact ($F \cdot t = m \cdot v$) on the Rocket Engine:** A GPS satellite has a mass of 2032 kg. To correct its trajectory, its velocity is to be increased by 0.1 m/s. The rocket engine used for this purpose is a power shock. The rocket engine used for this purpose ejects gas with a force of 100 N.

    a) What momentum must be transferred to the rocket?

    b) How long must the gas flow out to transfer this momentum?

    c) What momentum does the gas receive and how many kg of gas is required in total when it exits the nozzle at 150 m/s?

    d) The fuel consumption calculated in c) is too high. What would have to be changed to reduce the fuel consumption of a rocket?

5.8 Exercises

1. **Power, Energy and Efficiency of a Passenger Car:** A passenger car (m=1400 kg) accelerates uniformly with a = 2m/s² for 20s.

   a) What speed does the car reach?

   b) What is the kinetic energy of the car then?

   c) What is the acceleration power in kW (friction is neglected)?

   - Calculate the average power during acceleration.

   - Calculate the instantaneous power at the end of acceleration.

   d) Efficiency: The motor has an efficiency of 25%.

   - Calculate the energy that must be supplied to the motor during the acceleration phase.

   - Calculate the energy that the motor releases to the environment as heat during the acceleration phase.

   e) Law of conservation of energy: The engine is switched off and the car rolls up a slope at idle speed. How many meters up does it roll?

2. **Internal Energy in Inelastic Collision:** In a rear-end collision, vehicle A (m_A=?) collides with vehicle B (m_B=3500 kg). Both vehicles get caught in each other and continue to roll. In the collision, 60% of the kinetic energy is converted into internal energy (deformation and heat).

   a) State the law of conservation of energy.

   b) State the formula for calculating the velocity after an inelastic collision.

   c) Link the solutions a) and b) in a meaningful way, so that you can calculate the mass mA (mass vehicle A).

3. **Throw and Momentum:** From the shore of a lake a stone is thrown horizontally towards the lake (v_0= 20 m/s). The water surface is 6 m below the throwing position.

   a) How long does the stone fall?

   b) Name the two velocities v_x and v_y at the moment of impact on the water surface.

   c) What is the total velocity at impact?

   d) The stone has a mass of 0.6 kg and lands in a rowboat (30 kg).

   - Is this an inelastic or an elastic impact?

   - What velocity do the stone and the boat have in common if the boat was stationary before?

4. **Position Energy in a Chain Block (Pulley Block):** The chain block shown on
   the right is supposed to lift a bag (m_1 = 50 kg).

   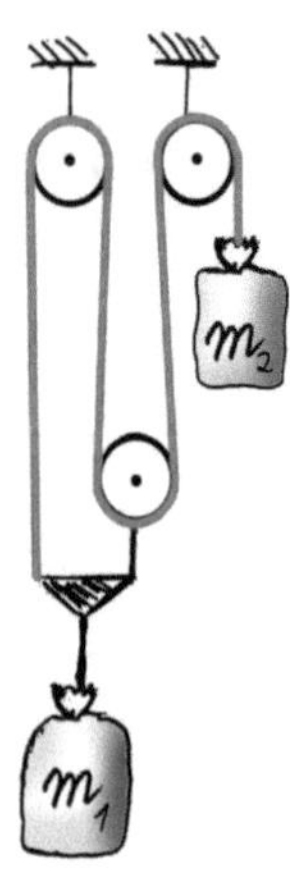

   a) What weight force would the mass m2 have to have in order for the sack m_1
      to move straight up?

   b) The following is now valid: m_2 = 40 kg. The mass is released. Does it move
      downward or upward?

   c) Calculate the difference in position energy (of both masses) when the mass
      m_2 (40 kg) has moved 5m.

   d) What are the velocities of both masses now?

5. **Acceleration of the Pulley:** Now calculate task 4d) again by first averaging the acceleration of
   mass 1 using the formula $\sum F_i = m \cdot a$. The following applies: $a_2 = 3\,a_1$. Then calculate
   $v_1 = \sqrt{2 \cdot a_1 \cdot s_1} = \dots$.

6. **Throw and Energy:** In an experiment, two stones of the same weight are thrown with a
   velocity of 20 m/s. The stones A and B are thrown vertically upwards. Stone A is thrown
   vertically upwards (90°). Stone B is thrown upward at an angle (30° to the horizontal).

   a) Both stones are thrown from a height of 2m. Calculate the time it takes for each to hit the
      ground.

   b) With which velocity (v_x, v_y and v_{total}) do the stones hit the ground?

   c) Using the law of conservation of energy, explain why both stones must have the same
      velocity when they hit the ground.

   d) What is the distance between the two stones after they have been in the air for 1s?

7. **Elastic Collision:** A vehicle collides with another
   vehicle. They collide perfectly elastically.

   a) Calculate the velocities after the collision.

   b) In this case, are the velocity differences before
      the impact and after the impact identical?

   c) Check whether the velocity differences before and after an elastic impact are basically
      identical.

8. **Tension Energy of a Spring:** A heavy metal ball (m=1kg) is placed on a spring which is
   compressed by 0.1 m (rest position). The spring is then tensioned by a further 0.4 m. What is the
   maximum height of the ball when it is released? (Let height 0 be the rest position). How fast is
   the ball when it just leaves the spring?

6 Kinematics of Rotation

6.1 Formula Symbols, Terms and the Angular Velocity

angular velocity	Winkelgeschwindigkeit	radian	das Bogenmaß
angular acceleration	Winkelbeschleunigung	the unit radian	die Einheit Radiant
orbital velocity	Bahngeschwindigkeit	the degree measure	das Gradmaß
path velocity	Bahngeschwindigkeit	period duration, periodic time	Periodendauer

■ Formula Symbols and Units

Translation (parallel displacement of a body)	Rotation

Kinematics

- path s   in m
- velocity v   in $\dfrac{m}{s}$
- acceleration a   in $\dfrac{m}{s^2}$

- angle φ   in rad
- angular velocity ω   in $\dfrac{rad}{s}$
- angular acceleration α   in $\dfrac{rad}{s^2}$

■ The Radian

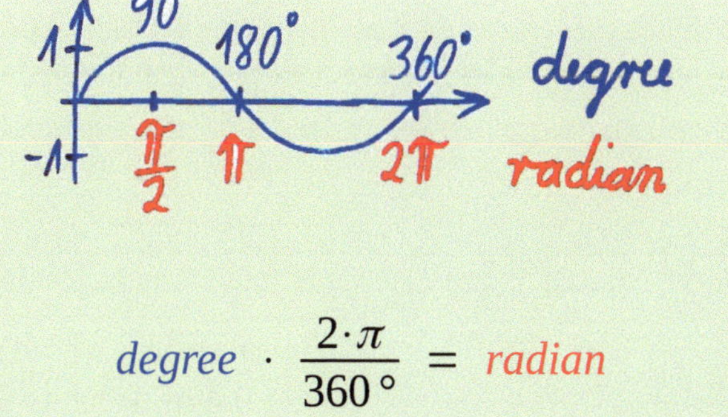

$$degree \cdot \frac{2 \cdot \pi}{360\,°} = radian$$

Note: If an angle is given with a degree sign (0°; 45°; 23.47°...) then set the calculator to degrees (**DEG**). If an angle is specified without the degree sign or with "rad" (0; π/2; 2.783 rad; ...) then set the calculator to radian (**RAD**).

■ The Angular Velocity "Omega"

Angular velocity

$$\omega = 2 \cdot \pi \cdot f = \frac{2 \cdot \pi}{T} \quad in \quad \frac{angle}{time} = \frac{rad}{s}$$

f = frequency in Hz = s⁻¹

T = period time in s

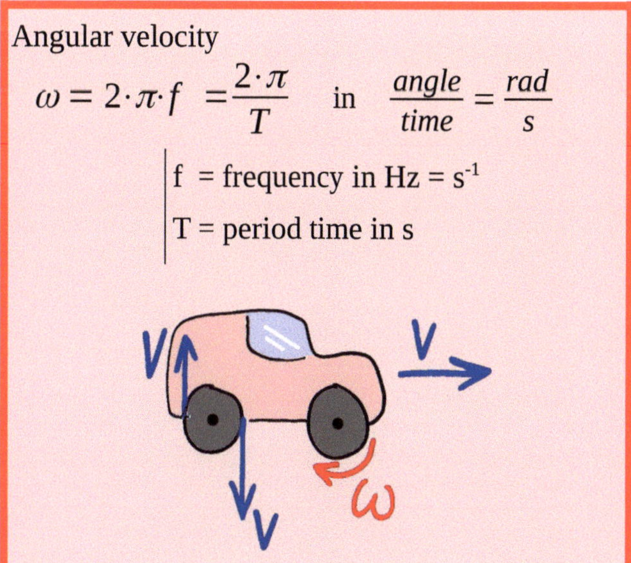

EXERCISES:

1. **Angular Velocity and Path Velocity:** Let the hour hand (the small hand) of a tower clock be 1.20 m long.

   a) What is its angular velocity ω (omega)?

   b) What is the orbital velocity at its tip?

2. **Angular Acceleration and Path Acceleration:** A car is driving on the highway at 130 km/h.

   a) What are the angular velocity ω and frequency f (=speed n) of its tires (d = 64cm)?

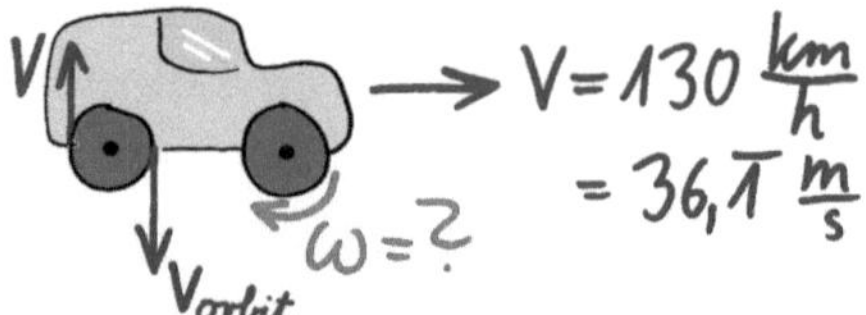

   b) The car brakes to 100 km/h within 10 seconds. What are the acceleration of the car and the angular acceleration of the tires?

 Badell.de

6.2 The Kinematics of Rotation - Compared to Translation

angular velocity	Winkelgeschwindigkeit	path, distance	der Weg
angular acceleration	Winkelbeschleunigung	screw press	die Spindelpresse
orbital velocity	Bahngeschwindigkeit	drive = frequency	Drehzahl = frequenz
path speed	Bahngeschwindigkeit	number of turns	die Anzahl der Umdrehungen
perimeter	Umfang	number of revolutions	die Anzahl der Umdrehungen
circumference	Kreisumfang, Ellipsenumfang		
circumferential force	Umfangskraft		

■ **The Linkage between Translation and Rotation**

Rolling ball: Linking the translation (s,v,a,F) with the rotation (φ,ω,α,M)

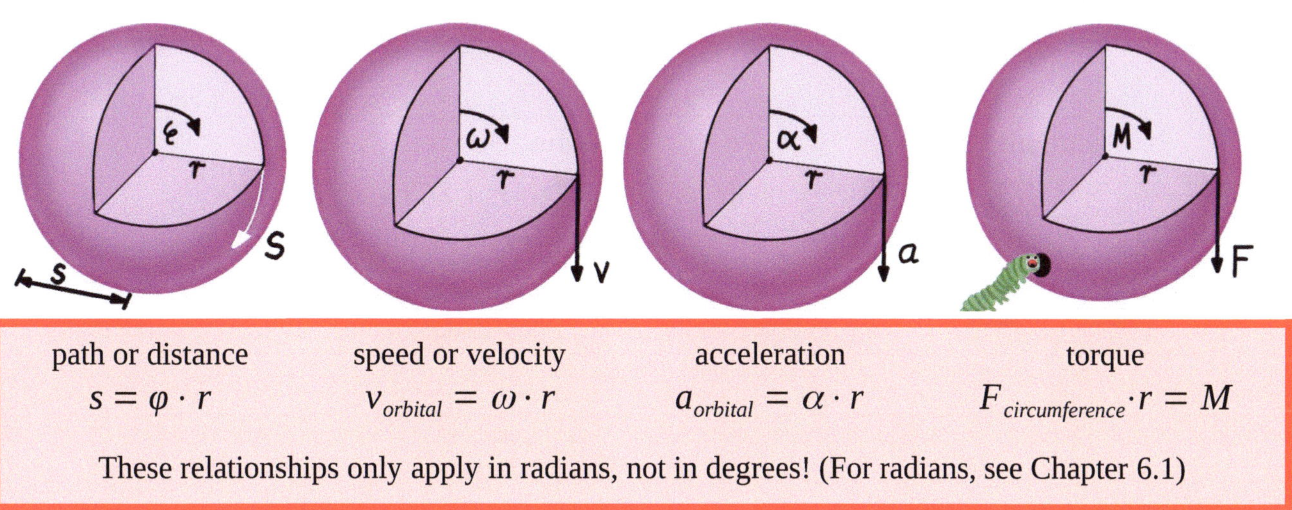

path or distance	speed or velocity	acceleration	torque
$s = \varphi \cdot r$	$v_{orbital} = \omega \cdot r$	$a_{orbital} = \alpha \cdot r$	$F_{circumference} \cdot r = M$

These relationships only apply in radians, not in degrees! (For radians, see Chapter 6.1)

A comparison of the formulas of translation and rotation shows that no new subject has to be understood here. On the contrary, all relations from the translation are directly transferable into the rotation. Only the variables have other names:

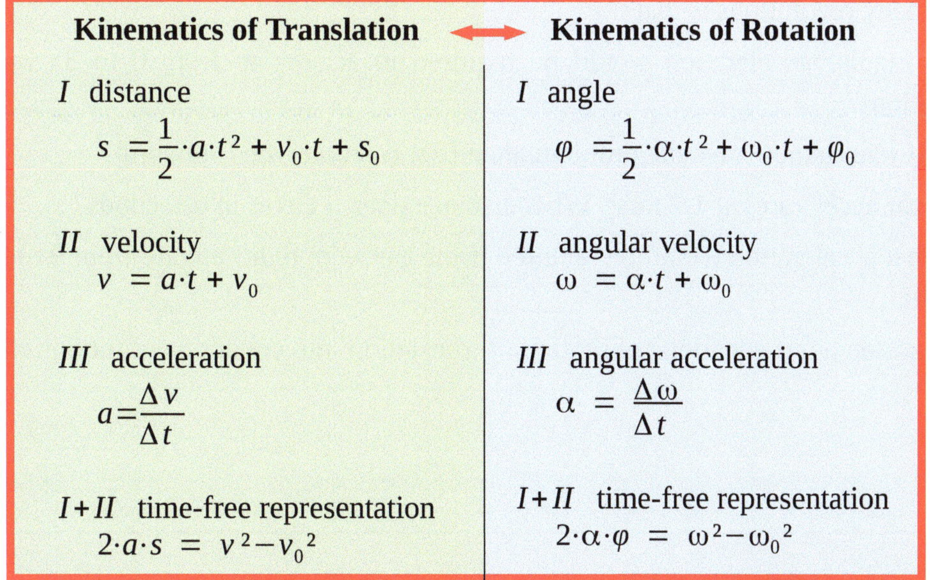

Kinematics of Translation ⟷ **Kinematics of Rotation**

I distance

$$s = \frac{1}{2} \cdot a \cdot t^2 + v_0 \cdot t + s_0$$

I angle

$$\varphi = \frac{1}{2} \cdot \alpha \cdot t^2 + \omega_0 \cdot t + \varphi_0$$

II velocity

$$v = a \cdot t + v_0$$

II angular velocity

$$\omega = \alpha \cdot t + \omega_0$$

III acceleration

$$a = \frac{\Delta v}{\Delta t}$$

III angular acceleration

$$\alpha = \frac{\Delta \omega}{\Delta t}$$

I + II time-free representation

$$2 \cdot a \cdot s = v^2 - v_0^2$$

I + II time-free representation

$$2 \cdot \alpha \cdot \varphi = \omega^2 - \omega_0^2$$

Table 6.1: Kinematics of Rotation

■ **EXAMPLE** (of Kinematics)

Kinematics of translation	**→ Kinematics of rotation**
Given: acceleration a = 3 m/s² time t = 8s v_0 = 0 m/s	**Given:** angular acceleration α = 4 $\dfrac{rad}{s^2}$ time t = 7s angular speed ω_0 = 0 $\dfrac{rad}{s}$
Sought: path s =?	**Sought:** **angel φ =?** **number of turns N =?**
Solution: $s = \dfrac{1}{2}\cdot a\cdot t^2 + v_0\cdot t + s_0 \ = 96\text{ m}$	**Solution:** $\varphi = \dfrac{1}{2}\cdot \alpha\cdot t^2 + \omega_0\cdot t + \varphi_0$ φ =98 rad (=5614°) N =98/2π =15,6 turns

EXERCISES

1. **Angular Acceleration of a Screw Press:** To optimize the production times of a screw press, a more powerful drive motor is to be used. This new motor achieves an angular acceleration of α = 1.2 rad/s² at the spindle.

   a) What angle does the spindle, which is initially stationary, cover within 6 seconds?

   b) How many revolutions is that?

   c) What angular acceleration would be required to accelerate from 0 to 3π rad/s within 5 revolutions?

   Hint: If you cannot solve the problem, then first calculate the following:

   - A car accelerates at 1.2 m/s². What distance does it cover in 6 seconds?

   - What acceleration would be required if the car were to accelerate from 0 to 3 m/s within 5m?

   - Translate your equations of motion of translation into equations of motion of rotation.

 ©*Badell.de*

6.3 Exercises

1. **Orbital Velocity on the Earth's Surface:** The earth (radius 6370 km) rotates once around its axis within 24 hours.

   a) With which orbital velocity does a person move forward who is standing at the equator?

   b) With which orbital velocity does a person move forward who is standing at the north pole?

   c) With which orbital velocity does a person move forward who is standing in the Ruhr area (latitude: 51.5° = angle between equator and Ruhr area)?

2. **Number of Revolutions during Braking:** A car drives so fast that the wheels turn 8 times per second. Within 40s the car brakes to a standstill (tire radius = 35 cm).

   a) What is the angular velocity of the tire and the orbital velocity of a point on the tire at the beginning of the braking process?

   b) Calculate the angular acceleration during the braking process.

   c) How many revolutions does the wheel make in total during the braking process?

3. **Orbital Velocity and Shaft Power:** Wind turbines must not exceed the speed of sound (340 m/s) with the ends of their blades. This would cause noise and mechanical overload of the blades. A 5 MW offshore wind turbine in the North Sea has a rotor diameter of 125 meters.

   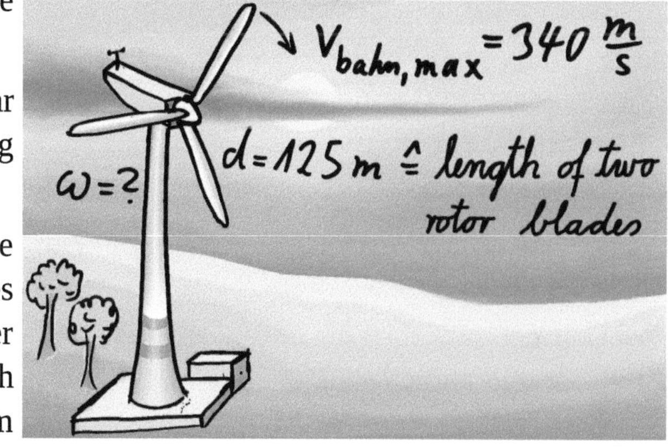

   a) Calculate the maximum possible angular velocity ω and the corresponding frequency f of the turbine.

   b) The blades rotate the shaft leading to the generator. The following formula applies to the power of a shaft: Power $P = \omega \cdot M$. Calculate the torque with which the shaft is rotated (maximum power at maximum speed).

   c) A common passenger car diesel engine has a torque of about 300 Nm (measured at the drive wheels). Compare the torques.

7 Dynamics of Rotation

rotational energy	Rotationsenergie	angular velocity	Winkelgeschwindigkeit
rotational power	Rotationsleistung	angular acceleration	Winkelbeschleunigung
(=shaft power)	(=Wellenleistung)	centrifugal force	Zentrifugalkraft (Fliehkraft)
angular momentum	Rotationsimpuls	centripetal force	Zentripetalkraft
torque, ,moment of force	Drehmoment	coriolis force	Corioliskraft
(mass) moment of inertia	(Massen-) Trägheitsmoment		
(geometrical) moment of inertia	(Flächen-) Trägheitsmoment		

While kinematics describes only motions and accelerations, dynamics also considers forces (or torques) and inertial masses (or moments of inertia).

7.1 Terms and Variables

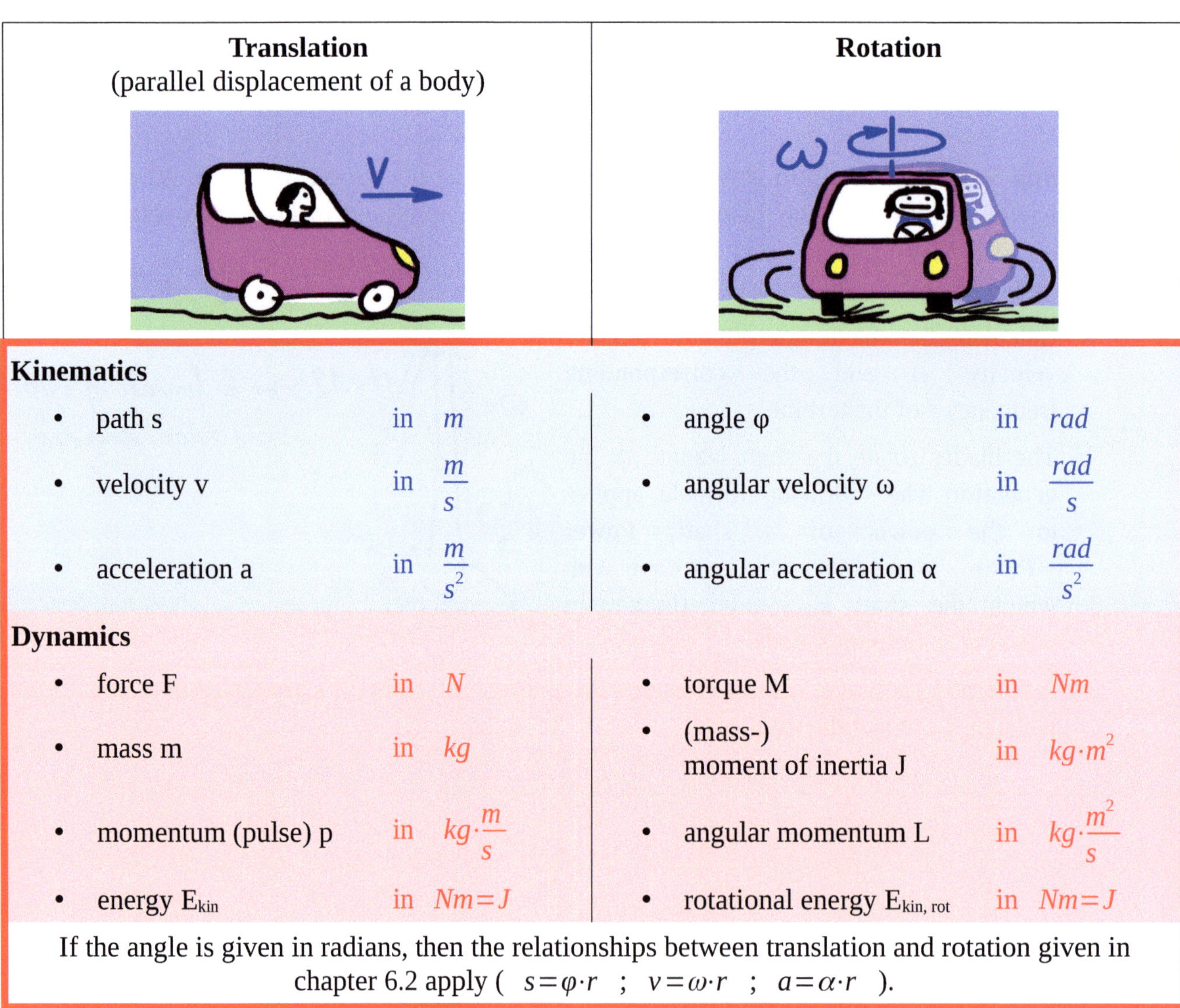

Translation (parallel displacement of a body)				Rotation			
Kinematics							
• path s	in	m		• angle φ	in	rad	
• velocity v	in	$\dfrac{m}{s}$		• angular velocity ω	in	$\dfrac{rad}{s}$	
• acceleration a	in	$\dfrac{m}{s^2}$		• angular acceleration α	in	$\dfrac{rad}{s^2}$	
Dynamics							
• force F	in	N		• torque M	in	Nm	
• mass m	in	kg		• (mass-) moment of inertia J	in	$kg{\cdot}m^2$	
• momentum (pulse) p	in	$kg{\cdot}\dfrac{m}{s}$		• angular momentum L	in	$kg{\cdot}\dfrac{m^2}{s}$	
• energy E_{kin}	in	$Nm=J$		• rotational energy $E_{kin,\,rot}$	in	$Nm=J$	

If the angle is given in radians, then the relationships between translation and rotation given in chapter 6.2 apply ($s=\varphi{\cdot}r$; $v=\omega{\cdot}r$; $a=\alpha{\cdot}r$).

Table 7.1: Rotation (variables)

7.2 Formulas

torque, moment of force	das Drehmoment	centrifugal force	die Zentrifugalkraft
(mass) moment of inertia	das (Massen-) Trägheitsmoment	coriolis force	die Corioliskraft
angular momentum	der Drehimpuls	kinetic energy	die kinetische Energie
shaft power	die Wellenleistung	rotational energy	die Rotationsenergie
tractive power	die Zugleistung		

With the exception of the centrifugal force and the Coriolis force, no new formulas are given in this chapter. As the table shows, all formulas of the translation can be used also for the rotation. Simply the variables are exchanged (variables, see chapter 7.1).

Dynamics of Translation	→ Dynamics of Rotation	
$force \quad F = m \cdot a$	$torque \; M = J \cdot \alpha$	→ see chapter 7.4
	$centrifugal\, force \; F_c = \dfrac{m \cdot v^2}{r} = m \cdot \omega^2 \cdot r$	→ see chapter 7.5
	$coriolis\, force \; F_{cor} = 2 \cdot m \cdot \omega \cdot v$	→ see chapter 7.6
$energy \quad E_{kin,trans} = \dfrac{1}{2} \cdot m \cdot v^2$	$energy \quad E_{kin,rot} = \dfrac{1}{2} \cdot J \cdot \omega^2$	→ see chapter 7.7
$momentum \quad p = m \cdot v = F \cdot t$	$angular\, momentum \; L = J \cdot \omega = M \cdot t$	→ see chapter 7.7
$tractive\, power \; P = F \cdot v \; = F \cdot \dot{s} = \dot{E}$	$shaft\, power \quad P = M \cdot \omega$	→ see chapter 7.8

Table 7.2: Dynamics of Rotation

7.3 The Torque

Translation: The Force	Rotation: The Torque	
$force \quad F \;$ in N	$torque$ $\quad M = F \cdot r \;$ in Nm	F = force r = radius

Exercises on torque can be found in volume 1 (Statics).

7.4 The (Mass-) Moment of Inertia J

(mass) moment of inertia	das (Massen-) Trägheitsmoment J	sphere	Kugel
rotational inertia	die Rotationsträgheit	steel shaft	Stahlwelle
(area) moment of inertia	das (Flächen-) Trägheitsmoment I	solid cylinder	Vollzylinder
second moment of area	das (Flächen-) Trägheitsmoment I	hollow shaft	Hohlwelle
bending stiffness	die Biegesteifigkeit	hollow cylinder	Hohlzylinder
angular acceleration	die Winkelbeschleunigung	dumb-ass	Vollpfosten
number of turns	die Anzahl der Umdrehungen	(Hint: Second moment is the moment of 2nd order. The order relates to the exponent → see also moments of a distribution in Vol.5: Statistics 1).	
revolutions, turns, rotations	Umdrehungen		

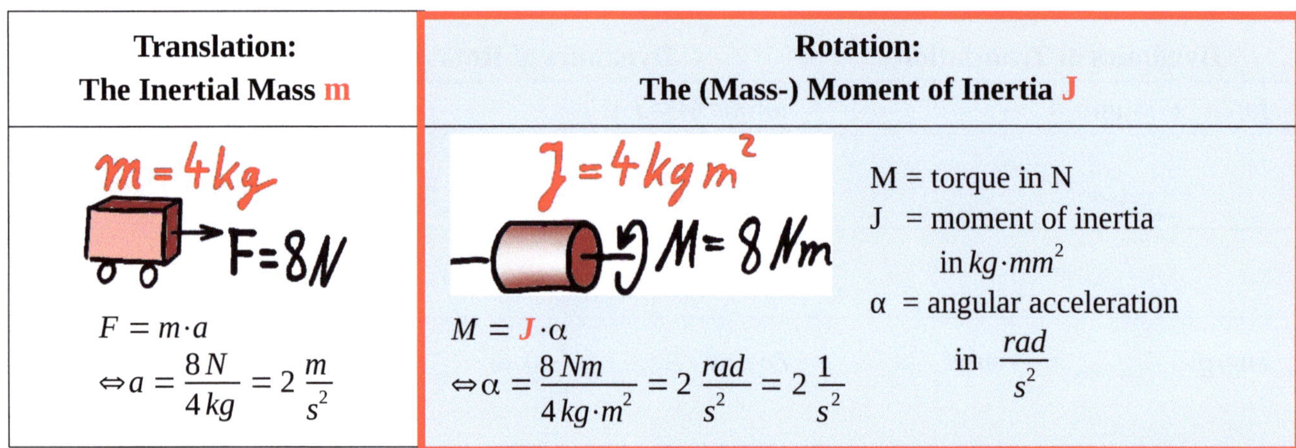

Translation:
The Inertial Mass m

$$F = m \cdot a$$
$$\Leftrightarrow a = \frac{8\,N}{4\,kg} = 2\,\frac{m}{s^2}$$

Rotation:
The (Mass-) Moment of Inertia J

$$M = J \cdot \alpha$$
$$\Leftrightarrow \alpha = \frac{8\,Nm}{4\,kg \cdot m^2} = 2\,\frac{rad}{s^2} = 2\,\frac{1}{s^2}$$

M = torque in N

J = moment of inertia in $kg \cdot mm^2$

α = angular acceleration in $\frac{rad}{s^2}$

Table 7.3: Moment of Inertia

The Mass Moment of Inertia is defined as follows:			
body	**characte-ristic**	**formula**	**definition**
	difficult to rotate	$J = m_1 \cdot r_1^2 + m_2 \cdot r_2^2$	The mass moment of inertia is defined as the sum of all mass points multiplied by their squared distances from the axis:
	easy to rotate	$J = m_1 \cdot r_1^2 + m_2 \cdot r_2^2$	
		$J = m \cdot r^2$	$J = \sum r_i^2 \cdot m_i = \int r^2 \cdot dm = \int r^2 \cdot \rho \cdot dV$

The following table shows solutions of the integral for various bodies. In this volume, only these ready-solved integrals are to be used. More information about the solution of the integral can be found in Volume 6: Mathematics in Physical Applications.

Caution: The <u>mass moment of inertia J</u> (rotational inertia) should not be confused with the <u>area moment of inertia</u> I or second moment of area (bending inertia). More about this in the volume "Engineering Mechanics".

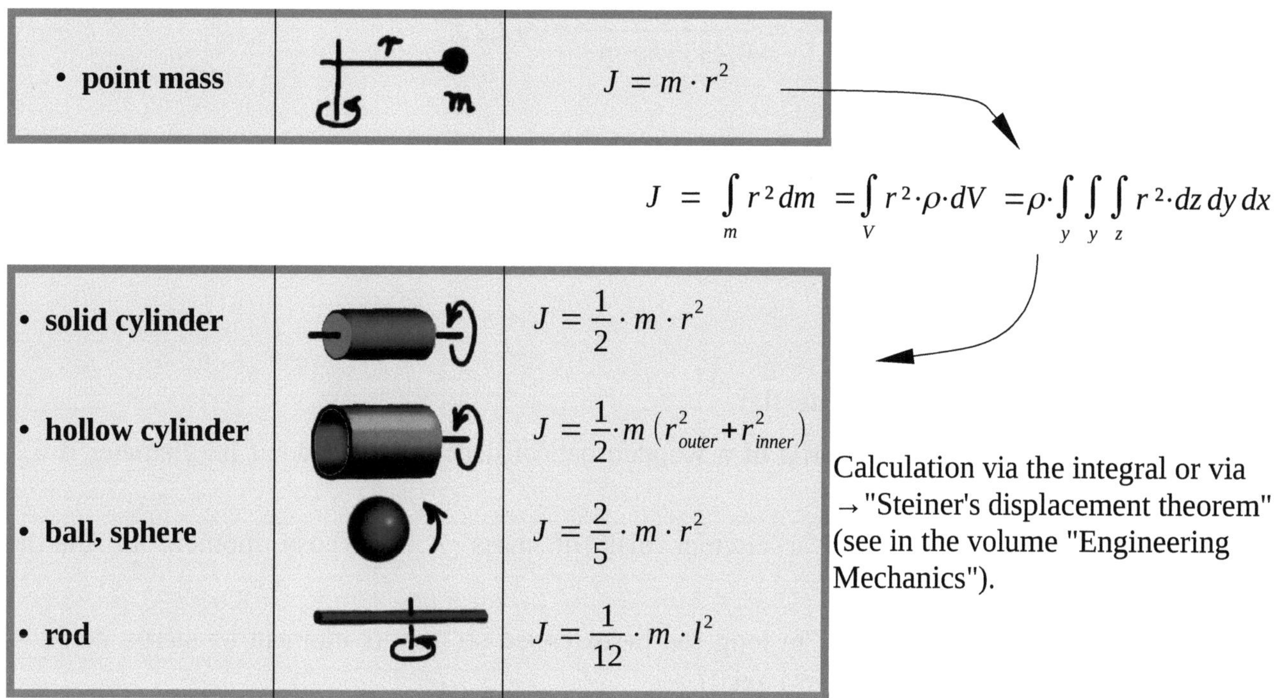

$$J = \int_m r^2\, dm = \int_V r^2 \cdot \rho \cdot dV = \rho \cdot \int_y \int_y \int_z r^2 \cdot dz\, dy\, dx$$

Calculation via the integral or via
→ "Steiner's displacement theorem"
(see in the volume "Engineering Mechanics").

Example 1

A sphere (solid ball) and a solid cylinder have the same mass (5 kg) and the same diameter (0.2 m). Which body can be rotated more easily?

Solution:

sphere: Mass moment of inertia $J_{ball} = \dfrac{2}{5} \cdot m \cdot r^2 = 0{,}02\,kg \cdot m^2$

Cylinder: Mass moment of inertia $J_{cylinder} = \dfrac{1}{2} \cdot m \cdot r^2 = 0{,}025\,kg \cdot m^2$

→ The ball is easier to turn than the cylinder.

Example 2

A steel shaft (density ρ = 8 kg /liter) has a diameter of 10 cm and a length of 3 m.

    a) What is its moment of inertia?

    b) What is its angular acceleration when rotated with a torque of 50 Nm (with no load at the other end)?

    c) What angle and how many revolutions does it make in the first 2 seconds?

Solution:

a) In the collection of formulas you will find the shaft under the term "cylinder".

$$m_{shaft} = \rho \cdot V_{shaft} = \rho \cdot r^2 \cdot \pi \cdot l = 188,5\,kg \qquad J_{shaft} = \frac{1}{2} \cdot m \cdot r^2 = 0,2356\,kg \cdot m^2$$

b) Angular acceleration:

$$[F = m \cdot a \rightarrow]\ M = J \cdot \alpha \ \Leftrightarrow\ \alpha = \frac{M}{J} = \frac{50\,Nm}{0,2356\,kg \cdot m^2} = 212,2\,s^{-2}$$

c) Total angle and number of turns:

$$[s = \frac{1}{2} \cdot a \cdot t^2 \rightarrow]\ \varphi = \frac{1}{2} \cdot \alpha \cdot t^2 = 424,4\,rad\ ;\qquad N = \frac{\varphi}{2 \cdot \pi} = 67,5\ \ revolutions$$

EXERCISES:

1. **Calculate some Moments of Inertia:**

   a) What is the moment of inertia of a wooden ball of density 700g/liter if its diameter is 12 cm?

   b) What is the diameter of a circular disk of mass 7 kg whose moment of inertia is $1,8\,kg \cdot m^2$?

   c) By what length l must a 0.7m long rod be extended so that its moment of inertia doubles (here: rotation about transverse axis)?

   d) What (kinetic) rotational energy do the bodies from task a) and b) each have when they rotate with a frequency of 10 s⁻¹?

2. **Torque at the Gyroscope:** With which torque M must a gyroscope with a moment of inertia J = 0.05 kg m² be driven, so that it reaches the speed 2000 min⁻¹ in 20s?

3. **Angular Acceleration of a Ferris Wheel:** A Ferris wheel is 30m high (d=30m). It is supposed to reach an orbital velocity of 5 m/s during operation.

   a) What is then the angular velocity ω of the Ferris wheel?

   b) What is the angular acceleration α if the Ferris wheel reaches its terminal velocity from a standing position within 15s?

   The Ferris wheel, in order to accelerate, is driven by a motor that directly turns the axis of the Ferris wheel.

   c) With what torque must the motor turn the Ferris wheel to achieve the angular acceleration α calculated above? The Ferris wheel has a moment of inertia of $J = 2 \cdot 10^6\ kg \cdot m^2$. Friction is neglected.

   d) What is the rotational kinetic energy E_{rot} of the Ferris wheel as it turns?

4. **Variable Moment of Inertia during a Pirouette:** A figure skater wants to do a pirouette. To do this, she takes momentum with her arms. As soon as she turns, she tightens her arms to rotate even faster.

   a) Why does the skater rotate faster when she tightens her arms?

   b) Roughly speaking, the figure skater (65 kg) has the shape of a vertical cylinder with a diameter of 30 cm and a height of 1.60 m. The figure skater's arms are in the shape of a cylinder. What moment of inertia does she have and what rotational energy does she reach when she spins very fast (3 revolutions per second).

   c) After the pirouette, she extends her arms again. This increases the mass moment of inertia to 270% of the previous value. How large is her rotational speed now? And what is the rotational kinetic energy?

5. **Variable Moment of Inertia during Speed Measurement:** The historical centrifugal governor of a steam engine consists of two balls rotating around an axis (principle according to James Watt, see picture). The greater the speed, the further the balls move away from the axis. The axis is assumed to be massless, the balls are point masses.

   The balls are suspended on 5 cm long rods. For a ball mass of mk = 80g, calculate the mass moment of inertia as a function of the angle (rod axis).

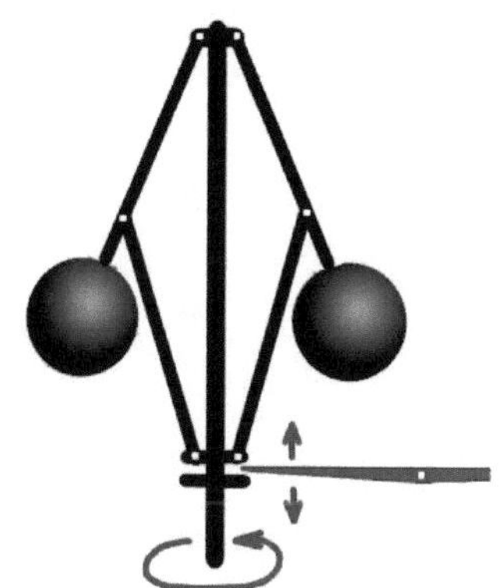

→ Further tasks see "Translation and rotation gekoppelt", chapter 8.

7.5 The Centrifugal Force F_c

weight (force)	die Gewichtskraft	steep turn	die Steilkurve
normal force	die Normalkraft	the car slides	das Fahrzeug rutscht
slope down force	die Hangabtriebskraft	the car tilts, the car tips over	das Fahrzeug kippt
radial force, central force	die Radialkraft	center of gravity	der Schwerpunkt
centrifugal force	die Zentrifugalkraft	tilting point (tipping point)	der Kipppunkt
centripetal force	die Zentripetalkraft	torque (torsional moment)	das Drehmoment
intertial force	die Trägheitskraft	tilting moment (overturning moment)	das Kippmoment M_K
acceleration force	die Beschleunigungskraft	stability moment	das Standmoment M_S
fictitious force	die Scheinkraft	stability	die Standsicherheit S
ficticious	fingiert		

■ **Radial Forces (Centrifugal Force and Centripetal Force)**

On a rotating body, the centrifugal force acts outward. The centripetal force (e.g. rope force or gravitational force) pulls the body inward.

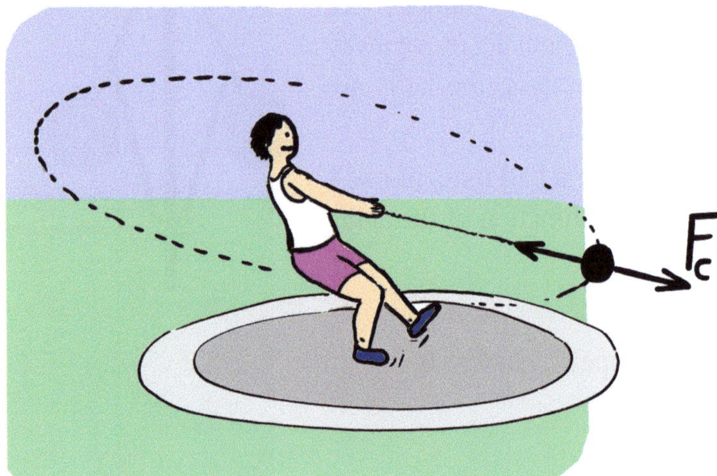

Radial forces F_r:

- The **centrifugal force** F_c (or F_{cf}) acts outwards.

- The **centripetal force** F_c (or F_{cp}), here as rope force, acts inwards.

  Centrifugal force and centripetal force are the same value.

centrifugal force	$F_c = \dfrac{m v^2}{r}$ $= m \cdot \omega^2 \cdot r$	centrifugal acceleration	$a_c = \dfrac{F_c}{m} = \dfrac{v^2}{r}$ $= \omega^2 \cdot r$

Table 7.4: Centrifugal Force

■ **The Centrifugal Force in a Loop**

In the loop, it should be noted that the centrifugal force always points outward, while the gravitational force always points downward. The total force on the vehicle $(\vec{F}_{total} = \vec{F}_c + \vec{F}_g)$ must therefore be determined differently at each position. While the forces add up at the bottom, the difference is formed at the top.

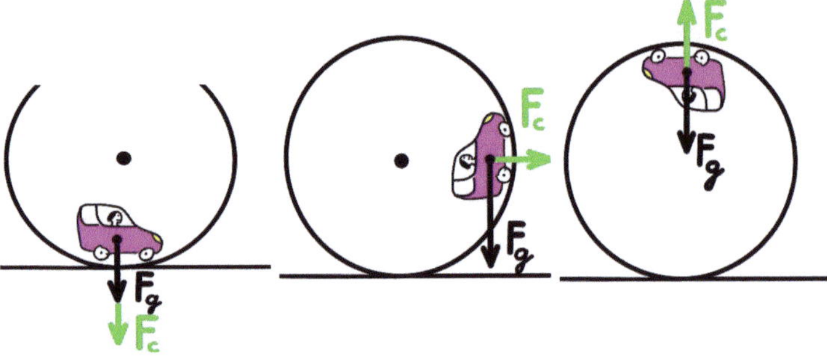

■ **The Centrifugal Force of Vehicles in a Curve:**

In the curve, the centrifugal force F_c acts outward, while the tire friction $(F_f = \mu \cdot F_n = \mu \cdot m \cdot g)$ acts inward. Tipping, on the other hand, involves a comparison of the torques between centrifugal force times center of gravity height and weight force times distance between center of gravity and tire.

vehicle slides	vehicle tilts	vehicle slides in a steep turn
F_C = centrifugal force F_f = frictional force F_n = normal force F_g = weight force	F_C = centrifugal force M_t = tilting moment $= l_1 \cdot F_z$ M_S = stability moment $= l_2 \cdot F_G$	F_C = centrifugal force $F_{C,T}$ = tangential component of F_C F_f = frictional force $= \mu \cdot F_n$ F_n = normal force $\quad = F_g \cdot \cos(\alpha) + F_c \cdot \sin(\alpha)$ F_h = downhill force
the vehicle slides, if $F_C > F_f$	the vehicle tips over, if $M_t > M_S$ → Stability see Volume "Engineering Mechanics"	the vehicle slides, if $F_{C,T} > F_f + F_h$

■ **The Centrifugal Force when walking**

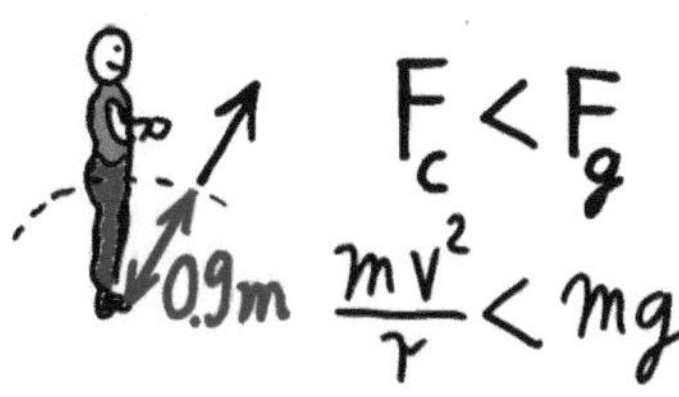

Even when walking, we are subject to centrifugal forces, since our hips describe a circular motion (see picture). Therefore, the maximum possible walking speed depends on the leg length. (If you want to get faster, you have to jog/run or bend your knees).

Example: What is the maximum walking speed of a person who has legs 90 cm long?

Solution: v=2.97 m/s = 10.7 km/h

- **Writing Styles - Everyone does it differently.**

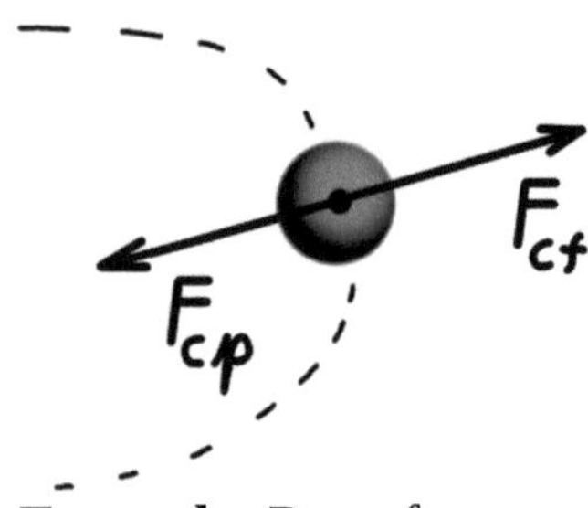

Example: Rope force as centripetal force.

In engineering mechanics, the equilibrium of forces would be set up for the radial direction. The positive direction could be defined as: "+" = outward. According to Newton I (4.2) then applies:

$$\sum F_{i,radial} = 0$$
$$\Leftrightarrow -F_{cp} + F_{cf} = 0$$
$$\Leftrightarrow F_{cf} = F_{cp}$$
$$\Leftrightarrow m \cdot \frac{v^2}{r} = F_{rope}$$

Actually, the centrifugal force is an inertial force, because the body is accelerated towards the center of the circle. From this point of view, one would use Newton II (4.2):

$$\sum F_{i,radial} = m \cdot a_c$$
$$\Leftrightarrow F_{cp} = m \cdot a_c$$
$$\Leftrightarrow F_{cp} = m \cdot \frac{v^2}{r}$$
$$\Leftrightarrow F_{rope} = m \cdot \frac{v^2}{r}$$

Quite trivially it can also be said:
Centrifugal force and centripetal force (=rope force) are equal:

$$F_{cp} = F_{rope}$$
$$m \cdot \frac{v^2}{r} = F_{rope}$$

- **The Term "Fictitious Force"**

The centrifugal force is an acceleration force or inertia force. It is sometimes also called "Fictitious force". I avoid this term. Because it creates the idea with pupils that the Fictitious force would accelerate only apparently. The same is true for the Coriolis force (reduce orbital velocity when moving to the center of rotation, chapter 7.6).

The origin of the term "apparent force" lies in the observation of a non-inertial (=rotating or accelerated) reference system, in which the observer knows nothing about the rotation/acceleration. For example, one could take a car as a reference system and then wonder by what mysterious, imaginary force one is pressed into the seat belt when a red light passes outside. The choice of a meaningful reference system avoids this problem.

EXERCISES:

1. **Law of Conservation of Energy in a Loop:** A vehicle goes through a loop. What is the significance of the law of conservation of energy for the vehicle in the loop? Describe the energy conversions.

2. **Survival in a Loop: A Stunt in a Loop is planned:** A passenger car driver is to drive through a loop at high speed. His goal is to survive: Thus, he should neither be crushed by too large forces, nor fall off the track due to too low speed.

   The first planning is: $m_{car} = 1,5$ t; $v_0 = 120$ km/h; $d_{loop} = 8$m.

   a) What is the speed of the car at the upper point of the loop?

   b) Does the car fall down?

   c) What is the largest force on the driver occurring in the loop (m = 80kg)?

   d) Would the driver survive this ride? He can withstand ten times his body weight.

 ®Badelt.de

3. **Design of the Wings of an Airplane:** To design a sports airplane suitable for aerobatics, the forces on the wings occurring during operation must be calculated. The airplane weighs 4 tons and should be able to fly through a loop (d = 60m) at a speed of v = 250 km/h.

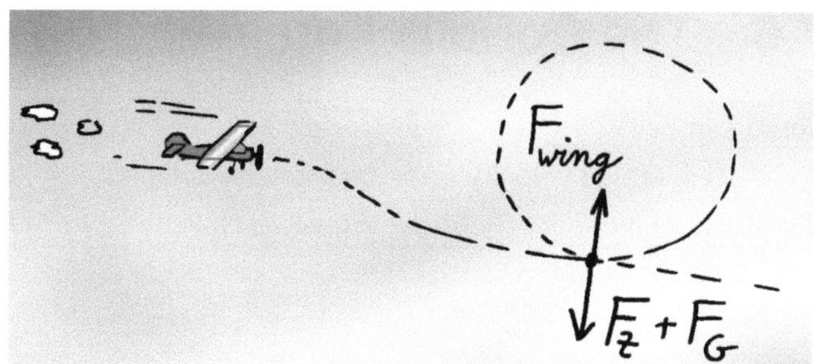

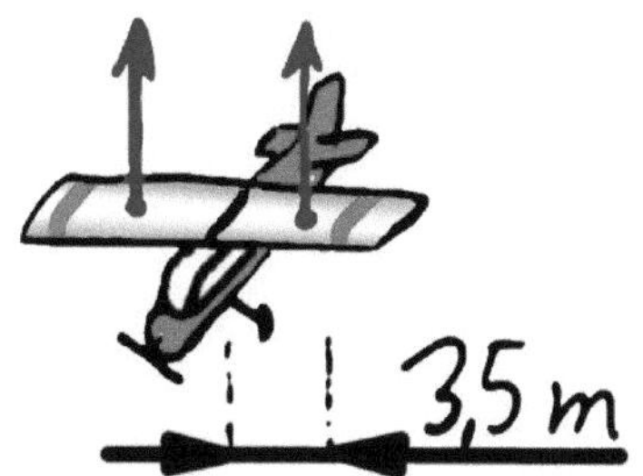

a) What is the maximum force (including weight force) acting on the wings?

b) The right wing is loaded centrally with half the force. The wingspan of the wings is 14m. What torque acts on the suspension of the wing?

4. **Centrifugal Force in a Curve:** A university lecturer drives home after class. At his home he drives his Sono Sion (m = 1400 kg) at a speed of 70 km/h into the highway exit. The task is to calculate whether the car can tip or slide away in the curve. The curve radius of the freeway exit is 50 meters.

a) Calculate the centrifugal force in the curve. Will the vehicle be braked by the centrifugal force?

b) Will the vehicle slide out of the curve? (Coefficient of static friction: tire on dry road $\mu = 0.8$; tire on wet road: $\mu = 0.5$.)

c) The vehicle weighs 1.4 tons. Its center of gravity is 0.2m above the ground. The car is 1.60 m wide. Will it tip over in the curve if it does not slide?

d) To allow higher speeds, the road is to be raised in the outside curve. What would be the minimum angle of inclination α for the car not to slip out of the curve on a wet road? 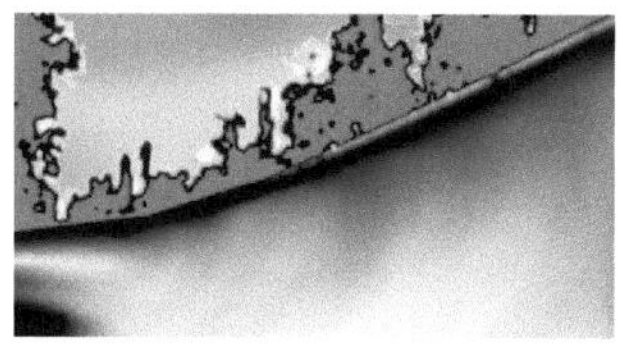

5. **Sliding and Tipping in a Curve:** A car (track width 1.3 m; center of gravity height 0.5 m) drives around a curve.

a) What is the smallest radius of curvature of the curve so that the car can drive through at 50 km/h without tipping over?

b) A curve with a radius of curvature of 30 m is now given. The tires have a static friction coefficient of $\mu_0 = 0.3$. How fast can the car drive around the curve without skidding?

7.6 The Coriolis Force

coriolis force	die Corioliskraft	detivative, derivation f'(x)	die Ableitung f'(x)
path speed, orbital velocity	die Bahngeschwindigkeit	derivation of a formula	die Herleitung einer Formel
angular velocity	die Winkelgeschwindigket	octopus arm	der Krakenarm

■ **Qualitative**

When running in the direction of the center of the circle, **the runner** must also decelerate at the same time: In the center, the path speed (v_{inner}) is smaller than outside (v_{outer}). For the running person it looks as if an unknown force accelerates him to the right.

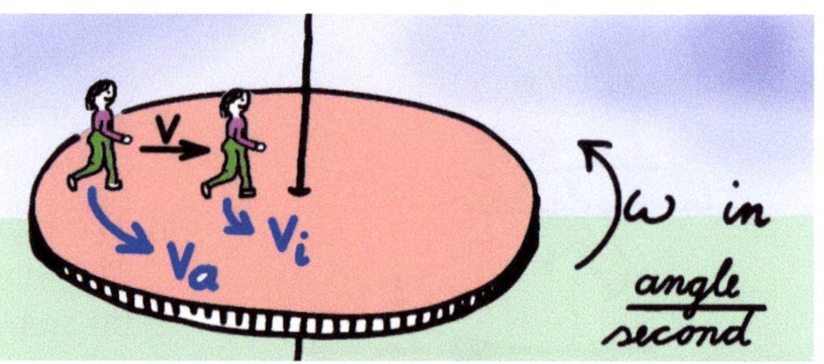

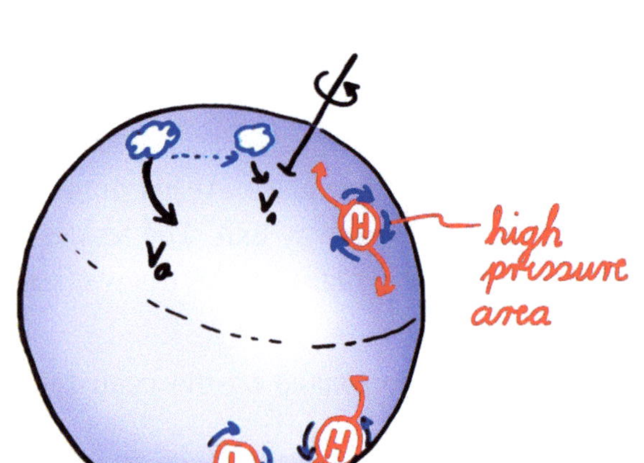

The wandering cloud on the globe on the left behaves exactly like the runner. It also experiences an immaginary acceleration to the right when the cloud moves to the axis.

Air flows outward from the high pressure area above. This air is also deflected. While high pressure areas in the northern hemisphere turn to the right, they turn to the left in the southern hemisphere. Low-pressure areas turn in the opposite direction.

> **Coriolis force** = inertia force for braking the path speed when moving to the center of rotation (or vice versa when moving outwards).
>
> If someone does not know that he is in a rotating system, then this Coriolis force is a virtual force (ficticious force) for him.

■ **Quantitative**

When running towards the axis of rotation, the path speed must be reduced. The following formulas apply to the calculation of the resulting virtual force:

Coriolis force:	v_r = radial velocity in the rotating system. (The cross product automatically takes into account the radial component, see „Mathematics in easy", Volume 4: Geometry).
$$F_C = 2m\omega v_r \qquad oder \qquad \vec{F}_C = -2m\vec{\omega}\times\vec{v}$$ Coriolis acceleration: $$a_C = \frac{F_C}{m} = 2\omega v_r \quad or \quad \vec{a}_C = -2\vec{\omega}\times\vec{v}$$	

Table 7.5: Coriolis Force

EXERCISES:

1. **Flight to Mars with artificial Gravity:** a flight to Mars takes about 6 to 9 months, depending on the position of the planets. Two space ships are to be connected with a 500 m long rope in such a way that they can orbit around each other during the flight to Mars.

   a) How large must the angular velocity of this rotation be, so that an artificial gravity (centrifugal acceleration) of 9.81 m/s² arises on the space ships? How long does one rotation take?

   b) An elevator is attached to the rope, which connects both spaceships. Calculate the Coriolis force on an astronaut (80 kg) for an elevator speed of v = 0.5m/s.

   c) Why does this Coriolis force occur and in which direction does it act?

2. **Coriolis Force in an "Octopus":** At a funfair, there is a ride called "The Octopus". Around the octopus head in the center, three octopus arms (main arms) rotate in a circle. At the ends of these arms are other octopus arms (secondary arms), which also rotate and hold the passenger gondolas. For the planning and construction of the ride, all the forces acting (centrifugal forces, acceleration forces, Coriolis forces...) must be known. Only then can the required strength of the beams, bearings, shafts and bolts used be calculated ($\rightarrow$ see also volume "Engineering Mechanics").

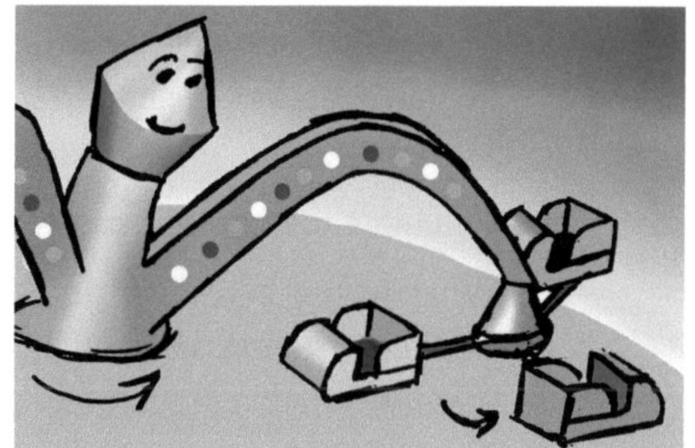

   Data: The main arms rotate at a maximum of 6 revolutions per minute. A gondola has a maximum mass of 500 kg including the two passengers. It moves with a maximum speed of v_{radial} = 10 km/h (orbital speed of the small circle) from the outside to the inside.

   a) With what angular velocity ω do the main arms of the ride move?

   b) What are the Coriolis force F_C and Coriolis acceleration a_C acting on the gondolas?

   c) The ride (main arms and side arms) has a diameter of 16 m, the side arms have a diameter of 4 m.

   - What is the maximum path speed of the gondolas caused by the main arms and the side arms together?

   - At which position do the largest centrifugal forces act on a nacelle? What are the maximum centrifugal forces acting on a nacelle?

   d) Additional task (Differential calculation): The maximum force F_{max}(centrifugal force + Coriolis force) is decisive for the strength calculation of the bolts. For the secondary arms, let φ be the angle by which the nacelle has moved away from the outermost point.

   - Set up a function for $F_{coriolis}(\varphi)$.

   - Set up a function for $F_{centrifugal}(\varphi) = F_{cf,\ main\ arm}(\varphi) + F_{cf,\ secondary\ arm}(\varphi)$.

   - Calculate the maximum force that occurs (derivative = 0).

3. **Coriolis Force in a High Pressure Area:** A weather report is always created using the "finite element method". Here, a component to be calculated - in this case clouds and air layers - is divided into countless, tiny volume elements. A very powerful computer then calculates which parameters (temperature, forces, pressure, flow...) are passed from each volume element to its neighboring elements.

   To calculate the rotation of a high pressure area, a finite element software also needs the Coriolis force that a volume element exerts on the neighboring elements.

   a) Air flows toward the north with an average velocity of 40 km/h.

      - If Germany is at latitude 51, what are the radial component and tangential component of this velocity?

      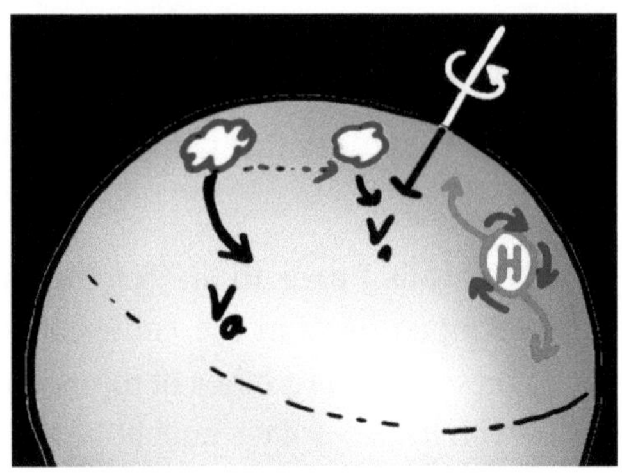

      - The software divides the air into volume elements of $100\,m \cdot 100\,m \cdot 100\,m$ each. What is the mass of such an air element if it consists of nitrogen, assuming simplified? ($N_2 = 2 \cdot 14.007$ g/mol; one mol has 22.4 liters).

   b) Calculate the Coriolis force acting on a volume element in our weather software.

   c) What would be the Coriolis acceleration of such a cloud if it could accelerate unimpeded (without friction with other air masses)? Compare the result with the acceleration due to gravity: What is the ratio a_c / g_{Earth}?

4. **Coriolis Force on a Raindrop:** A raindrop (d = 2mm) falls on the earth at the equator.

   a) Which weight force F_g acts on the drop? What velocity of fall v does it reach? (Stokes friction force for spheres in laminar flow: $F_f = 6\pi \cdot \eta \cdot r \cdot v$ with viscosity $\eta_{air} = 17{,}1\,\mu Pas$)

   b) What Coriolis force F_{cor} does it experience? (Water weighs 1 kg per liter).

   c) At what angle would it fall to earth in absolute still air?

7.7 Energy and Momentum of Rotation

kinetic energy	die kinetische Energie	(mass) moment of inertia	das (Massen-) Trägheitsmoment J
translational energy	die Translationsenergie	rotational inertia	die Rotationsträgheit
rotational energy	die Rotationsenergie	(geometrical) moment of inertia	das (Flächen-) Trägheitsmoment I
pulse, momentum	der Impuls	bending stiffness	die Biegesteifigkeit
rotational pulse	der Rotationsimpuls	see engineering mechanics	s. Technische Mechanik
force of impact	der Kraftstoß		

■ The Energy of Rotation

The translation energy (kinet. energy of translation)	The Rotational Energy (kinetical energy of rotation)	
$E_{kin,trans} = E_{kin} = \frac{1}{2} \cdot m \cdot v^2$	$E_{kin,rot} = E_{rot} = \frac{1}{2} \cdot J \cdot \omega^2$	J = (mass-) moment of inertia ω = angular velocity
Mass **m**: How hard is it to push?	Mass moment of inertia **J**: How hard is it to turn?	
Velocity **v**: How fast does it go?	Angular velocity **ω**: How fast is it spinning?	

Table 7.6: Energy of Rotation

■ The Momentum of Rotation

Translational Momentum	Rotational Momentum	
momentum $\quad p = m \cdot v$	momentum $\quad L_{rot} = J \cdot \omega$	L = rotational momentum J = (mass-) moment of inertia ω = angular velocity
force of impact $\Delta p = F \cdot t$	force of impact $\Delta L_{rot} = M \cdot t$	

Table 7.7: Momentum or Rotation

EXERCISES

1. **Energy and Momentum of Rotation:** on a common shaft (m = 0 kg) there are two massive flywheels (approximately solid cylinders):

   m_1=10kg; d_1=0.6m and m_2=8kg; d_2=0.4m.

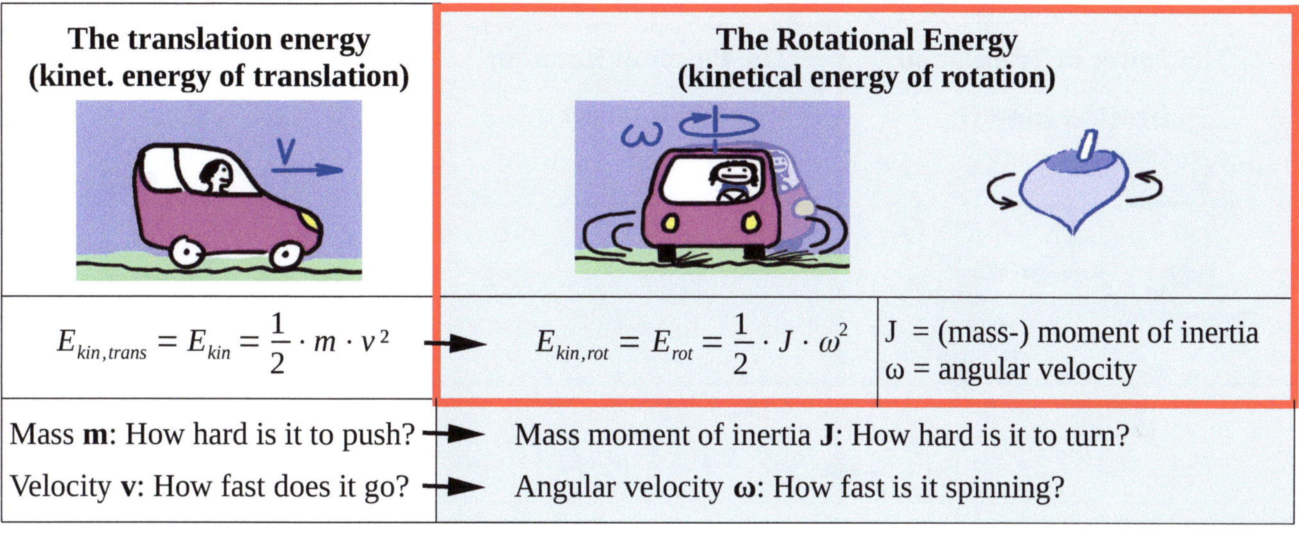

   a) Initially, disk 1 rotates at a speed n (=frequency f) of 100 per minute. What is its rotational energy?

   b) What is its rotational momentum?

   c) The second disk, which did not rotate before, is now clutched. Is this an elastic impact or a inelastic impact? Is rotational energy also converted into heat during the impact?

   d) What is the common rotational frequency of the two disks?

   e) How much heat energy is released?

7.8 The Power of Rotation (Shaft Power)

shaft (= solid cylinder)	Welle (= Vollzylinder)	gearbox	das Getriebe
shaft power	die Wellenleistung	cardan shaft	die Kardanwelle
rotational power	die Rotationsleistung	tension, tensile stress	die Zugspannung
torque, moment of force	das Drehmoment	shear stress	die Scherspannung
tractive power	die Zugleistung	shear stress	die Schubspannung
translational power	die Translationsleistung	bending stress	die Biegespannung
tractive force	die Zugkraft		

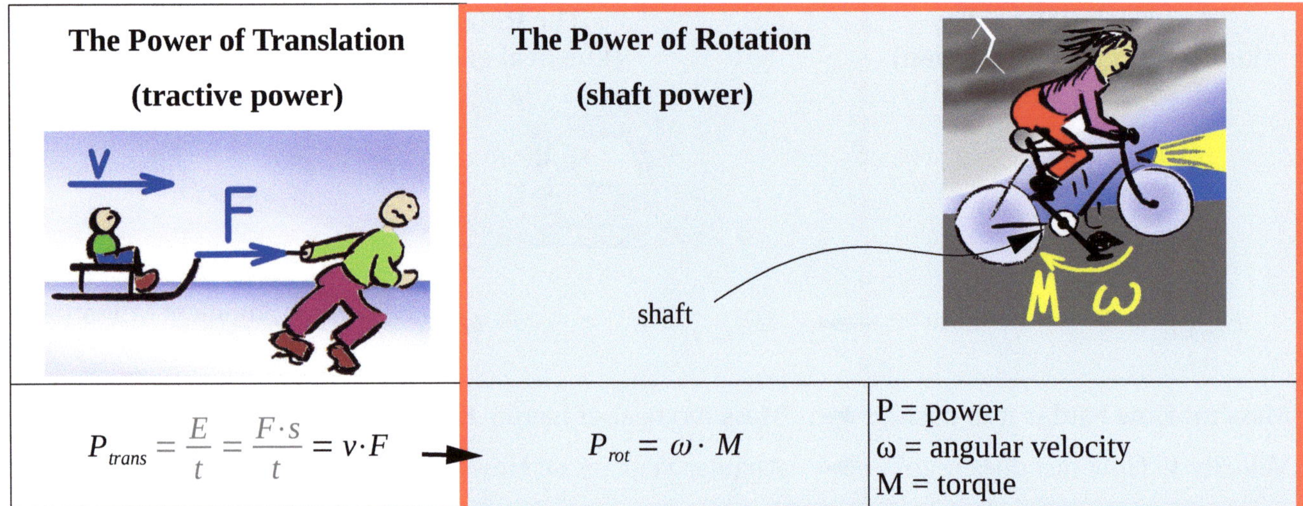

$$P_{trans} = \frac{E}{t} = \frac{F \cdot s}{t} = v \cdot F \qquad\qquad P_{rot} = \omega \cdot M$$

Table 7.8: Power of Rotation

EXERCISES

1. **Design of a Cardan Shaft:** A motor has a rated power of 80 kW at 3000 rpm. A shaft is to be purchased to connect this motor to the gearbox.

   a) What torque must the shaft withstand?

   b) Behind the gearbox (efficiency η =98%, transmission ratio i=6) the speed is lower. What torque must this shaft withstand?

   c) Additional task "Engineering Mechanics": The stress in materials is given in Newton per mm². Stresses are calculated with the following four formulas:

tensile stress (= tension)	shear stress (due to shearing)	bending stress	torsional stress (due to the torque)
$\sigma = \dfrac{F}{A}$	$\tau = \dfrac{F}{A}$	$\sigma_b = \dfrac{M_b}{W}$	$\tau_t = \dfrac{M_t}{W_t}$

   F = Force with which the shaft is pulled or sheared
   A = Shaft cross-sectional area
   M = torque (bending moment or torsional moment)
   W = section modulus = $\dfrac{\pi}{16} \cdot d^3$ for solid cylinders (shafts).

   What torsional stress would occur in the shaft from sub-task a) if its diameter was d=20mm? Would you use a steel grade that can withstand 300 N/mm²?

7.9 Exercises

1. **Toy Car falls out of the Loop:** A toy car (m=200g) drives at the point A with the speed v_A into a loop (d = 60cm). Friction is neglected.

   a) Equilibrium of forces: What velocity v_B does the car need at the top of the loop to avoid falling down?

   b) Law of conservation of energy: What is the minimum velocity v_A before looping if the car drives through the loop without falling down?

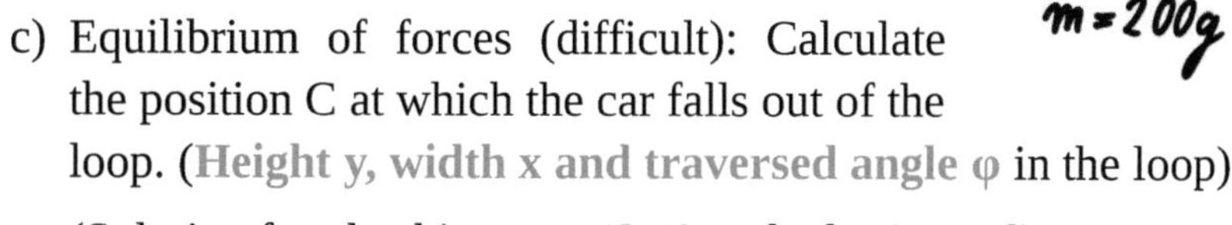

   In a second experiment, the car travels at only 90% of its previous initial speed.

   c) Equilibrium of forces (difficult): Calculate the position C at which the car falls out of the loop. (**Height y, width x and traversed angle** φ in the loop).

   (Solution for checking: φ =43.1° to the horizontal).

   d) What is the velocity v and its components v_x and v_y at the moment when the car leaves the track (point C)?

   The car now falls down.

   e) Oblique throw: From point C the car describes a throw parabola. How far does the car fly until it hits the ground (x-distance)?

   f) Oblique throw: At what angle does the car fall to the ground?

2. **Formula understanding:** Check the validity of the following formulas for rolling bodies.

   a) Newton's second law $F_{Zug} = m \cdot a$ with F_{zug} = tractive force or traction

   - Why is the formula not valid for rotating bodies?
   - How large is the error?
   - What is the correct formula? (Solution: $F_{Zug} - F_{Umfang} = ma \Leftrightarrow F_{Zug} = ma + \dfrac{J \cdot \omega}{r}$)

   b) Conservation of momentum (impact of two bodies): $m_1 \cdot v_1 + m_2 \cdot v_2 = m_1 \cdot u_1 + m_2 \cdot u_2$

3. **Tidal Friction shifts the Moon:** Graviations of the moon create ebb and flow on the earth. This tidal friction causes the Earth to spin 17 microseconds slower each day.

   a) Assumed, this would have always been so: How long would an Earth day have to have been 4 billion years ago (shortly after the Earth and Moon were formed)?

   b) We assume further, the earth is a halfway homogeneous ball. What is then its moment of inertia and what rotational energy does the earth emit daily?

The tidal friction does not only lead to frictional heat. It also has the consequence that the orbital velocity of the moon becomes faster: The ebb and flow of the tide are so time-delayed that the tidal wave accelerates the moon with its gravity. For the following calculations we assume that 50% of the emitted rotational energy is converted into frictional heat and 50% into motion energy of the moon.

   c) What are the current position energy and orbital velocity of the moon?

   d) Which energy is added to the moon in the course of one year?

   e) How large are position energy and orbital velocity of the moon in the next year?

   f) By how many centimeters does the moon move away from the earth annually?

8 Translation and Rotation coupled

8.1 The Rolling Body

the coupling of… / the link between...	die Verknüpfung von…/ die Kopplung von...
translational energy and rotational energy	Translationsenergie und Rotationsenergie
path, angle, circumference	Weg, Winkel, Umfang
velocity, angular velocity, orbital velocity	Geschwindigkeit, Winkelgeschwindigkeit, Bahngeschwindigkeit
acceleration, angular acceleration, orbital acceleration	Beschleunigung, Winkelbeschleunigung, Bahnbeschleunigung
force and torque are loads	Kraft und Drehmoment sind Lasten

■ **The Coupling of Translation and Rotation in a rolling Body**

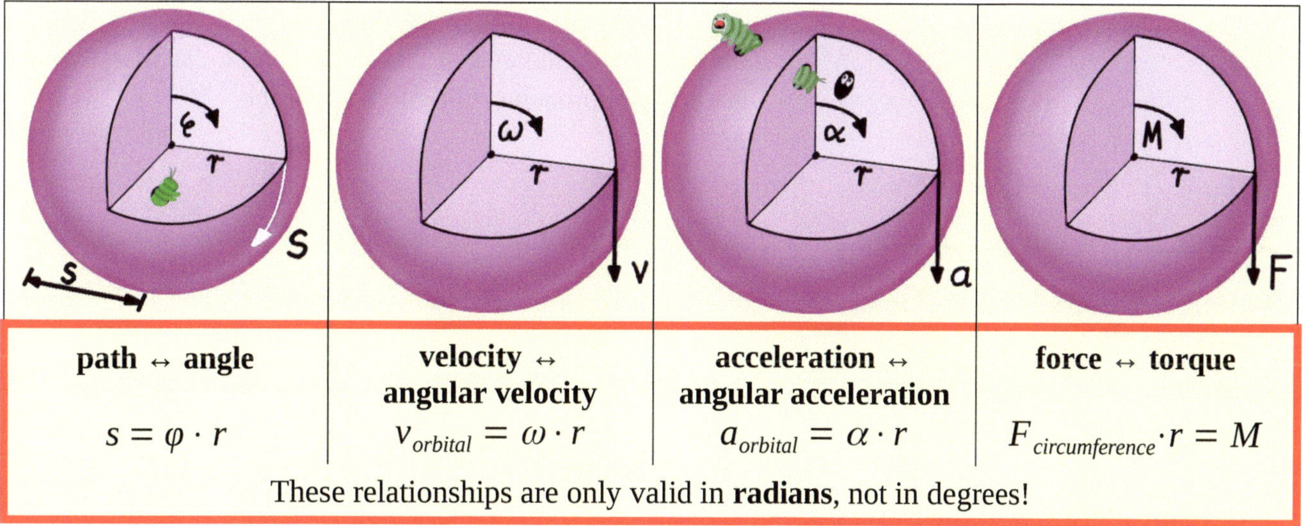

path ↔ angle	**velocity ↔ angular velocity**	**acceleration ↔ angular acceleration**	**force ↔ torque**
$s = \varphi \cdot r$	$v_{orbital} = \omega \cdot r$	$a_{orbital} = \alpha \cdot r$	$F_{circumference} \cdot r = M$

These relationships are only valid in **radians**, not in degrees!

Table 8.1: Linkage between Translation and Rotation

EXERCISES

1. **Rolling Distance of a Solid Ball:** A wooden ball (r = 4cm; m = 0.2 kg) rolls with the velocity v = 4m/s in a horizontal plane.

   a) What is the moment of inertia of the ball? (For formulas see chapter 7.4).

   b) Calculate the rotational energy and the translational energy.

   c) How far does the ball roll if the rolling friction is μ = 0.02?

2. **Rolling Distance of a Hollow Ball:** a ball (ρ=400g/l, d_a =300mm, d_i =295mm) is rolled with v_0 = 3m/s over a horizontal plane (friction coefficient μ = 0.04). There is no air resistance. How far does the ball roll?

   (Note: moment of inertia of a hollow ball = moment of inertia of the solid ball minus moment of inertia of the missing material).

8.2   The Rolling Body on the Slope

slope	der Hang	(mass-) moment of inertia	das (Massen-) Trägheitsmoment
weight (force)	die Gewichtskraft	energy approach	der Energieansatz
gravitational force	die Gravitationskraft	the law of conservation of energy	der Energieerhaltungssatz
slope down force	die Hangabtriebskraft	force approach	der Kraftansatz
tangential force	die Umfangskraft	sum of forces	die Kräftesumme
tangential force	die Tangentialkraft	inclined plane	die schiefe Ebene
potential energy	die Lageenergie	sphere	die Kugel
potential energy	die potentielle Energie	solid cylinder	der Vollzylinder
translational energie	die Translationsenergie		
rotational energy	die Rotationsenergie		

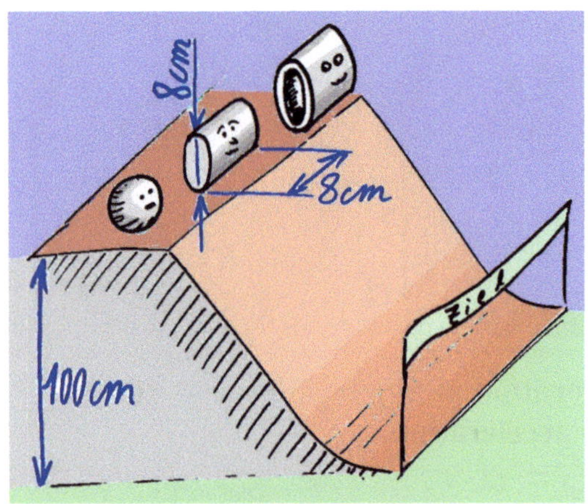

The velocity at the end of the slope can be calculated using the energy approach or the force approach.

■ **Consideration of the Forces**

A rolling body on a slope is

- accelerated by the slope down force
- braked by the mass moment of inertia

$$F_H = F_G \cdot \sin(\alpha) = m \cdot g \cdot \sin(\alpha)$$

$$F_{Umfang} \cdot r = M_{Trägheit} = J \cdot \alpha$$

■ **Consideration of the Energies**

From the position energy E_{pot}

the kinetic energies E_{trans} and E_{rot} are calculated.

$$Epot = m \cdot g \cdot h$$

$$E_{trans} = \frac{1}{2} \cdot m \cdot v^2 \quad ; \quad E_{rot} = \frac{1}{2} \cdot J \cdot \omega^2$$

The kinetic energy of translation (E_{kin} or E_{trans} or $E_{kin,trans}$) is also called translation energy.

The kinetic energy of rotation (E_{rot} or $E_{kin,rot}$) is also called rotational energy.

 Badell.de

The Speed at the End of the Slope

Calculation with the Energy Approach	Calculation with the Force Approach

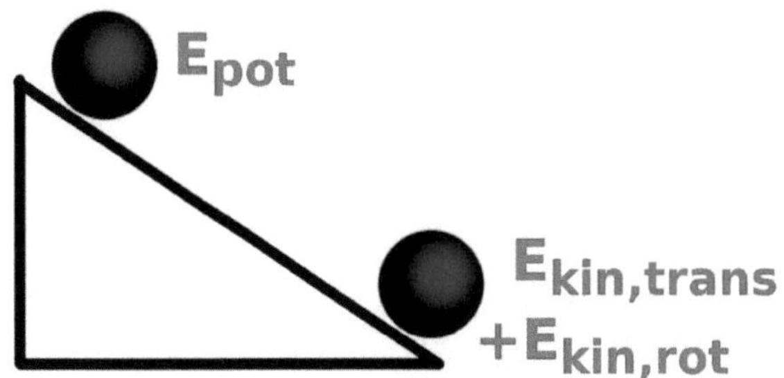

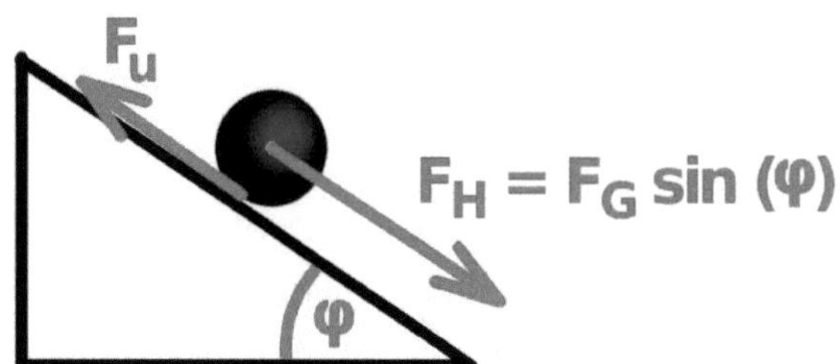

Calculation with the Energy Approach

$$E_{pot} = E_{kin} + E_{rot}$$

$$\Leftrightarrow m \cdot g \cdot h = \frac{1}{2} \cdot m \cdot v^2 + \frac{1}{2} \cdot J \cdot \omega^2 \qquad \Big| \ \omega = \frac{v}{r}$$

$$\Leftrightarrow \qquad = v^2 \cdot \left(\frac{1}{2} \cdot m + \frac{1}{2} \cdot J \cdot \frac{1}{r^2} \right)$$

$$\Leftrightarrow \qquad v = \ldots$$

Calculation with the Force Approach

Sum of forces:

$$\sum F_i = m \cdot a$$

slope down force:

$$F_H = F_G \cdot \sin(\phi) = m \cdot g \cdot \sin(\phi)$$

tangential force:

$$F_{Umfang} \cdot r = M_{Trägheit} = J \cdot \alpha$$

$$\Leftrightarrow F_{Umfang} = \frac{J \cdot \alpha}{r}$$

Inserting all forces into the equation:

$$m \cdot g \cdot \sin(\phi) - \frac{J \cdot \alpha}{r} = m \cdot a$$

$$\Big| \text{ coupling: } a = \alpha \cdot r \ \Leftrightarrow \ \alpha = \frac{a}{r}$$

$$\Rightarrow m \cdot g \cdot \sin(\phi) = \frac{J \cdot a}{r^2} + m \cdot a$$

$$\Leftrightarrow \frac{m \cdot g \cdot \sin(\phi)}{m + \dfrac{J}{r^2}} = a$$

After the acceleration was calculated, the speed can be calculated with the formula

$$2 \cdot a \cdot s = v^2 - v_0^2 \ .$$

8.3 Exercises

1. **Law of Conservation of Energy on the Inclined Plane I:** A <u>sphere</u>, a <u>solid cylinder</u> and a <u>hollow cylinder</u> roll down a ramp simultaneously. The ramp is 1m high and has a slope angle of 25°. All bodies have an outer diameter of d_{outer} = 8 cm and are 8 cm wide. The hollow cylinder has an inner diameter of d_{inner} = 6 cm.

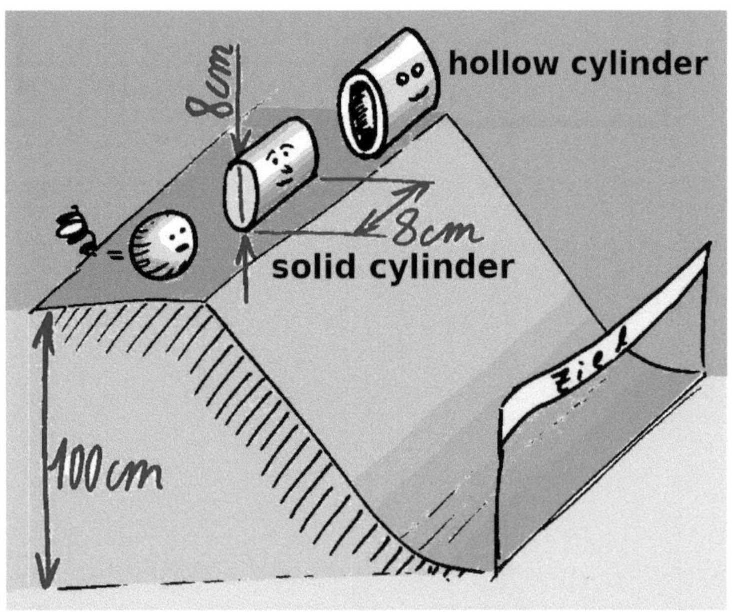

   a) The bodies are made of steel (8 kg per liter). Calculate the masses and the (mass) moments of inertia of all three bodies.

   b) Using the law of conservation of energy, calculate: What velocity do each of the bodies reach at the end of the ramp?

   c) What are the accelerations and the times required in each case (angle of incline 25°)?

   d) What is the ratio of the translational velocities between the solid cylinder and the sphere for the general case: diameter = width = d?

   e) Calculate the downhill force F_H for the sphere and justify why the equation $F_H = m \cdot a$ does not apply here.

2. **Law of Conservation of Energy on the Inclined Plane II:** A ball rolls down a 1m high ramp. Justify whether the velocity of the ball reached at the end becomes larger or smaller if you lengthen the ramp and thus reduce its inclination.

   a) For the case where there is no friction.

   b) For the case where rolling friction occurs.

3. **Rolling Ball in a Loop:** A ball rolls first down a slope (h=2m high) and then through a loop (r_L = 0.5m radius). The ball has a radius of r_{ball}=0.1m and a mass of 4 kg.

   Note: The (mass) moment of inertia of a sphere is $J = \dfrac{2}{5} m \cdot r^2$

   a) What is the moment of inertia of this sphere?

   b) What is the velocity of the ball after it rolls down the slope?

   c) The ball rolls through the looping track. What is the velocity of the ball at the top of the loop?

   d) Does the ball pass through the loop or does it fall down?

 © Badell.de

9 Gravitation

9.1 Homogeneous and Radial Gravitational Field

gravitational force	die Gravitationskraft	homogeneous field	das homogene Feld
gravitational field strength	die Gravitationsfeldstärke g	radial field	das Radialfeld
gravitational acceleration	die Erdbeschleunigung g	radial force	die Radialkraft
gravitational constant	die Gravitationskonstante G	centrifugal force	die Zentrifugalkraft (Fliehkraft)
Newtons laws	die Newtonschen Gesetze	centripetal force	die Zentripetalkraft

For the gravitational force was used so far: $F_G = m \cdot g$ $\quad$ with $g = 9{,}81\,\dfrac{N}{kg} = 9{,}81\,\dfrac{m}{s^2}$.

However, the gravitational field strength g is not constant at all. Its value 9.81 N/kg is valid exclusively on the earth surface in middle latitudes (Europe, USA, China).

In contrast, the gravitational constant G is generally valid. With its help, gravitational forces in the radial field, such as the attraction of two spheres or two planets on each other, can be described.

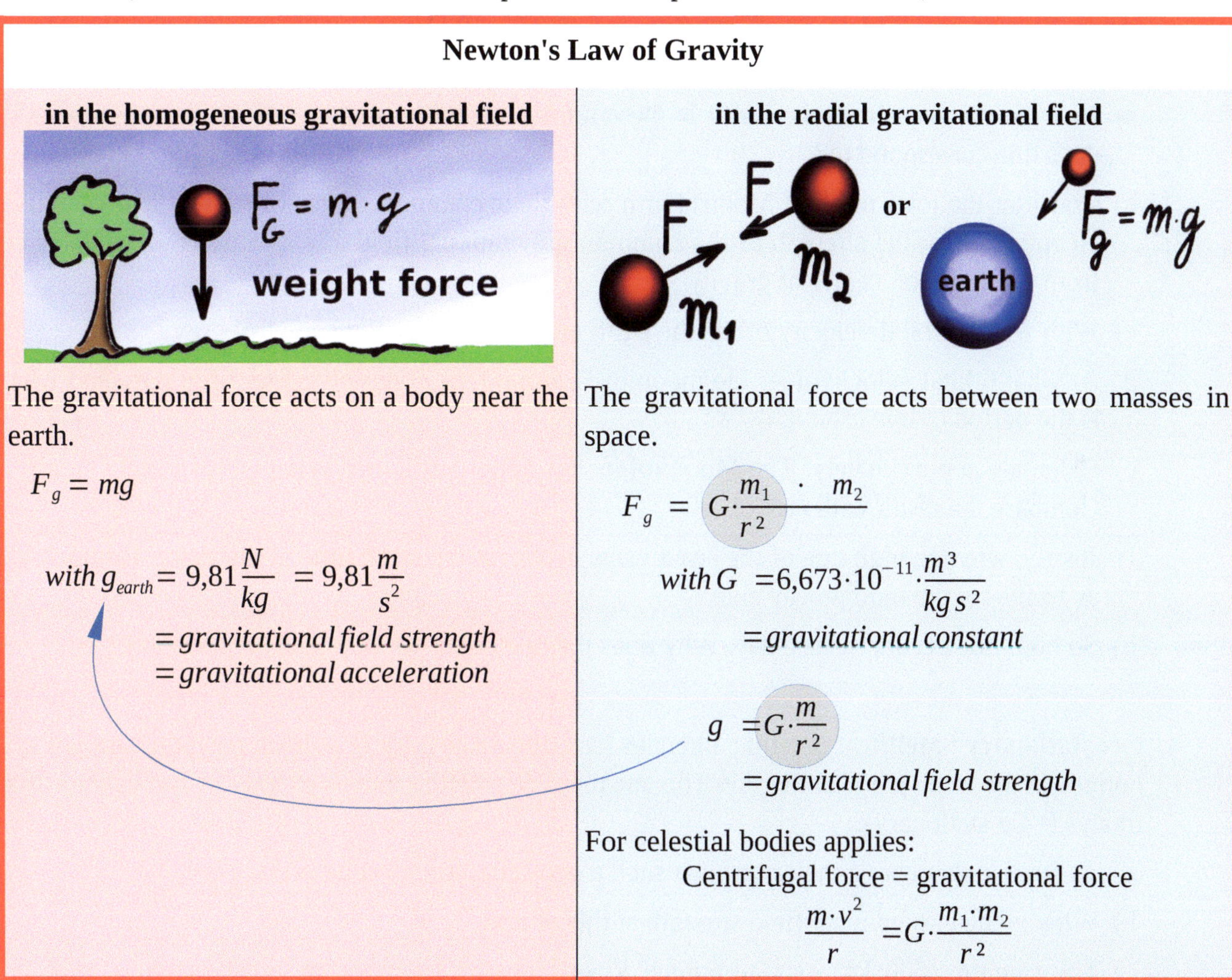

Table 9.1: Gravitational Field

EXERCISES

1. **Gravitational Field Strength and Gravitational Constant:** The gravitational constant $G = 6{,}673 \cdot 10^{-11} \frac{m^3}{kg \cdot s^2}$ is different from the gravitational field strength $g = 9{,}81 \frac{N}{kg}$.

    a) Explain the difference between g and G.

    b) Calculate the gravitatinos field strength g_{Earth} on the Earth's surface using the gravitational constant G.

    c) Calculate the gravitatinos field strength g_{Moon} on the lunar surface using the gravitational constant G.

2. **Solar Mass:** the mass of the earth is $5{,}97 \cdot 10^{24} kg$. Since the mass of the earth is negligible compared to the much larger mass of the sun, we can say that the earth orbits the sun. The distance earth-sun is 8 light minutes. What is the mass of the sun?

3. **Moon Phases and Tides:** The Earth (m_E = 5.9722 - 1024 kg; r_E = 6378 km) and the Moon (7.346 - 1022 kg) have a mean center-of-mass distance of 384,000 km.

    a) At what distance from the Earth is the common center of gravity? How many Earth radii does this correspond to?

    b) Consider the total mass of Moon+Earth with their common center of gravity and establish an equation for the strength of the common gravitational field g_{ME} as a function of distance from the common center of gravity.

    c) With which orbital velocity orbits the earth around the common center of gravity?

    d) At what orbital velocity does the moon orbit around the common center of gravity and what is the period of one orbit in days?

    e) Calculate approximately: The Moon orbits the stationary Earth. What is its orbital velocity? Compare the result with task c).

    f) Justify, why the high tide of the sea always occurs at the same time at the moon-facing side, as well as at the moon-away side.

    g) So high tide occurs twice a day. Why does the tide come 40 minutes later each day?

4. **Geostationary Satellites:** Satellite antennas (satellite dishes) for television reception are always pointed at the same place in the sky. This means: The satellite moves exactly as fast as the earth rotates (=geostationary).

    a) At what height above the earth must such a geostationary satellite orbit?

    b) What is the gravitational field strength at this altitude?

    c) The satellite (500 kg) gets additional kinetic energy by a rocket accelerating it. Will its angular velocity increase or decrease? Will its orbital velocity increase or decrease?

9.2 Kepler - Planetary Orbits

planetary orbit, orbit of a planet	die Planetenbahn	to the power of three	hoch drei
circular orbit, circular path	die Kreisbahn	to sweep	überstreichen
elliptical orbit, elliptical path	die elliptische Bahn	to sweep	kehren, fegen

The German astronomer, mathematician and theologian Johannes Kepler (1571-1630) discovered the laws according to which planets move around the sun. According to him, the planetary orbits behave according to the following laws:

1) All planets move on elliptical orbits. In one of the foci is the Sun (this law is approximately true, because the mass of the Sun can be considered very large compared to the other planetary masses).

2) A line drawn from the Sun to a planet sweeps equal areas in equal times. This means that planets near the sun move faster (have more kinetic energy), while far from the sun they have more positional energy and less velocity. This observation shows the validity of the law of conservation of energy also in the universe.

3) In his third law, Kepler established a proportionality between the orbital periods of the planets (to the power of 2) and the size of the elliptical orbit (length of the semi-axes of the ellipse to the power of 3).

9.3 The Pendulum

period duration	Periodendauer	The oscillation equation	die Schwingungsgleichung
frequency	Frequenz f	Spring pendulum	das Federpendel
drive	Drehzahl n	spring constant	die Federkonstante
angular velocity	die Winkelgeschwindigkeit	thread pendulum	das Fadenpendel
forms of energy	Energieformen	thread length	die Fadenlänge
law of conservation of energy	der Energieerhaltungssatz	stopped pendulum	das Hemmungspendel
		Galileo pendulum	das Hemmungspendel

- **The cause of an Oscillation** is always the change between two forms of energy (for details see volume 4: Oscillations, waves and optics).

Example spring pendulum:

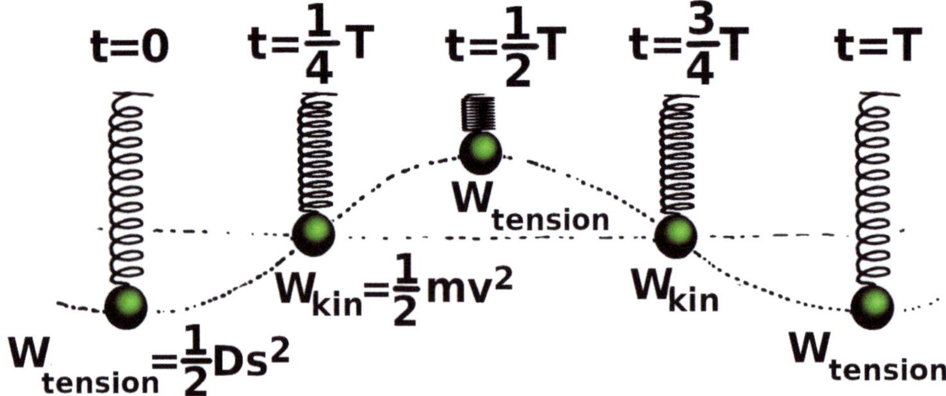

Period duration T in s; angular velocity ω in s^{-1}; spring constant D in N/m

- **Formulas for Spring Pendulum and Thread Pendulum**

	The Spring Pendulum	The Thread Pendulum
forms of energy	Tension energy of the spring and kinetic energy of the ball	Position energy in the gravitational field and kinetic energy of the ball
period duration	$T = 2\cdot\pi\cdot\sqrt{\dfrac{m}{D}}$	$T = 2\cdot\pi\cdot\sqrt{\dfrac{l}{g}}$
	$\omega = \dfrac{2\cdot\pi}{T} = \dfrac{\sqrt{D}}{m}$	$\omega = \dfrac{2\cdot\pi}{T} = \dfrac{\sqrt{g}}{l}$
	m = mass D = spring constant	l = thread length g = gravitational field strength
oscillation equation	$position\ y(t) = \hat{y}\cdot\sin(\omega\cdot t) = \hat{y}\cdot\sin\left(2\pi\cdot\dfrac{T}{t}\right)$ $velocity\ v(t) = \dot{y}(t)$	

Table 9.2: Spring Pendulum and Thread Pendulum

■ **Law of Conservation of Energy**

An oscillation can always occur when the available energy changes between two forms of energy. The sum of all energies is the same at any time, unless frictional heat is dissipated.

EXERCISES

1. **Spring Pendulum:** A spring hangs on a hook and is not deflected (position 0 m).

   a) After suspending a mass m_1 = 200g, the spring expands by 5 cm. Calculate the spring constant D.

   b) The spring is at rest with mass m_1. After attaching another mass m_2 = 150g, the spring starts to oscillate. Calculate the amplitude of oscillation and the position of the new rest position of the spring.

   c) Oscillation equation $s_y(t) = y(t) = \hat{y} \cdot \cos(\omega \cdot t)$ (Radiant! Chapter 6.1):

   - At what position is the spring after two seconds?

   - What is its velocity at that moment?

   - What is its acceleration at that moment?

2. **Design of a Clock Pendulum:** Today, clocks are built with a quartz. This is a tiny, vibrating crystal that looks like a tuning fork. In the past, clocks were built with pendulums. The rotating pendulum in a pocket watch (spiral spring with rotating mass) is called a balance. Large grandfather clocks, on the other hand, have an oscillating mass on a swinging rod. There are two types of pendulums:

   - Mathematical pendulum: Ideles thread pendulum with point-like mass.

   - Physical pendulum: Real pendulum with a given mass distribution.

   In the following task we assume a clock pendulum, which is simplified as a mathematical pendulum.

   a) A large grandfather clock is to be constructed in such a way that the clock pendulum has a period of 2 seconds. The pendulum is to be 1m long. Calculate how large the pendulum mass must be.

   b) The pendulum is deflected 15 degrees and then released. Calculate the maximum orbital velocity at the zero crossing.

9.4 Exercises

1. **Forces and Masses on the Moon:** An astronaut walks over the moon. He has a mass of 130 kg including luggage. The gravitational acceleration on the moon is 1.62 m/s². The moon radius is 1739 km (earth radius = 6370 km).

   a) What is the mass of the moon?

   b) With which force is the astronaut pushed onto the moon?

2. **Velocities of Satellites:** The Earth's radius is 6378 km, and the Earth's mass is $5{,}972 \cdot 10^{24}\,kg$. Answer the following questions about the speeds of satellites:

   a) A satellite is flying around the Earth at an altitude of 200 km. What is the speed of the satellite?

   b) A GPS satellite flies at an altitude of 20180 km above the Earth's surface. Calculate its orbital period around the Earth in hours.

3. **Altitude of a Satellite:** A news satellite is supposed to fly in a geostationary orbit. This means that it rotates exactly as fast as the Earth itself. The satellite has the mass m.

   a) At what altitude does the satellite fly above the earth?

   b) Does the altitude depend on the mass m?

4. **The second Kepler's Law (speed of a planet):**

   a) State Kepler's second law.

   b) Describe qualitatively how Kepler's second law is consistent with the law of conservation of energy.

5. **The Stopped Pendulum:** A stopped pendulum (Galileo pendulum) is a string pendulum whose string length is reduced on one side by an obstacle.

   The following quantities are given:

   - Thread length left: $l_1 = 15$ cm
   - Thread length right: $l_2 = 40$ cm.
   - Mass m = 40g

   a) Calculate the period T.

   b) Determine the ratio of the rise heights h_1 / h_2.

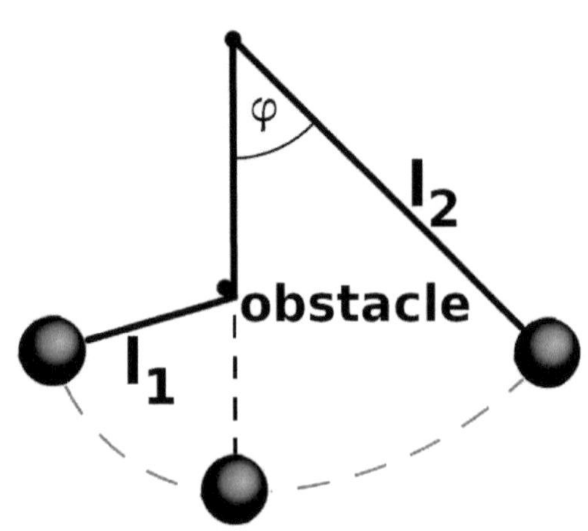

10 Appendix

10.1 Solutions

You will find a collection of handwritten solutions (photos from the whiteboard) on my website badelt.de. It should not be a problem that the solutions are in German, since they are formulas and calculations anyway.

10.2 List of Tables